**NASA
Reference
Publication
1301**

1993

Annular Solar Eclipse of 10 May 1994

Fred Espenak
Goddard Space Flight Center
Greenbelt, Maryland
USA

Jay Anderson
Prairie Weather Centre
Winnipeg, Manitoba
CANADA

National Aeronautics and
Space Administration

Office of Management

Scientific and Technical
Information Program

PREFACE

Since 1949, the U. S. Naval Observatory has published a special series of circulars containing detailed information for upcoming solar eclipses. These predictions were provided as a public service and were of vital importance to the international astronomical community in the planning and execution of successful eclipse expeditions. Unfortunately, the USNO *Circulars* were discontinued in 1991.

This has left a real and tangible void for detailed and accurate predictions for future solar eclipses. The information is not only of great interest and value to the scientific community in general, but to NASA in particular. For instance, Hubble Space Telescope passed through the Moon's shadow during the recent total solar eclipse of 11 July 1991. Without adequate advanced warning of this event, the eclipse would have had serious consequences on HST's energy budget in the rapidly diminishing sunlight. Furthermore, solar eclipses are known to have important effects on Earth's ionosphere and therefore play a significant role in the interaction and understanding of the Earth-Sun environment. Finally, NASA has a history of participating in various solar eclipse experiments through both ground based and aerial (i.e. - Kuiper Airborne Observatory, sounding rockets) investigations.

With the issuance of this NASA Reference Publication, the authors plan to continue the tradition of providing special bulletins containing extensive, detailed and accurate predictions and meteorological data for future solar eclipses of interest. The eclipse bulletins are provided as a public service to both the professional and lay communities, including educators and the media. In order to provide a reasonable lead time for planning purposes, subsequent NASA RP's for future eclipses will be published 18 to 24 months before each event. Single copies of these RP's will be available at no cost, provided a written request is received after publication. A special request form for the eclipse RP's may be found on the last page of this publication, and may be returned to Jay Anderson. Comments, suggestions, criticisms and corrections are solicited in order to improve the content and layout in subsequent editions of this publication series, and may be sent to Fred Espenak.

Permission is freely granted to reproduce any portion of this Reference Publication, including data, figures, maps, tables and text (except for material noted as having been published elsewhere, or by persons other than the authors). All uses and/or publication of this material should be accompanied by an appropriate acknowledgment of the source (e.g. - "Reprinted from *Annular Solar Eclipse of 10 May 1994*, Espenak and Anderson, 1993"). This work is in no way associated with or supported by the U. S. Naval Observatory or its *Circulars* series of publications.

Fred Espenak
NASA/Goddard Space Flight Center
Planetary Systems Branch, Code 693
Greenbelt, MD 20771
USA

Jay Anderson
Prairie Weather Centre
900-266 Graham Avenue
Winnipeg, MB,
CANADA R3C 3V4

Next NASA Eclipse RP :   *Total Solar Eclipse of 3 November 1994*
Available: *Autumn 1993*

ANNULAR SOLAR ECLIPSE OF 10 MAY 1994

Table of Contents

ANNULAR SOLAR ECLIPSE OF 10 MAY 1994

Figures, Tables and Maps

ECLIPSE PREDICTIONS

INTRODUCTION

On 10 May 1994, an annular eclipse of the Sun will be widely visible from the Western Hemisphere. The Moon's anti-umbral shadow delineates a path through North America including northern Mexico, the American southwest and midwest, the southern Great Lakes region, New England and maritime Canada. The path crosses the North Atlantic where it sweeps over the Azores and ends at sunset in Morocco. From within the much broader path of the Moon's penumbral shadow, a partial eclipse will be seen from North American, eastern Siberia, western Europe and north Africa (Figures 1 and 2).

PATH AND VISIBILITY

The path of the Moon's anti-umbral shadow begins in the Pacific Ocean about 700 kilometers southeast of the Hawaiian Islands. As the shadow first contacts Earth along the sunrise terminator (15:21 UT), the path is 311 kilometers wide and the annular eclipse lasts 4 minutes 34 seconds. Quickly sweeping east-northeast across the Pacific, the shadow's first landfall occurs at 15:54 UT along the western coast of Baja California (Figures 3 and 4). The path width has diminished to 267 kilometers as the curvature of Earth's surface brings subsequent points in the path closer to the vertex of the umbra. Nevertheless, the duration of annularity has increased to 5 minutes 28 seconds. This occurs because the surface or ground component of the velocity vector in the direction of the shadow's motion has increased fast enough to over compensate for the effects that the narrower path width has on the duration of annularity. From Baja, the Sun will appear 40° above the horizon and the eclipse magnitude will reach 0.939 at maximum eclipse. This corresponds to an obscuration of 0.882 of the total surface area of the Sun's disk. As the anti-umbra rushes across the Golfo de California it moves with a ground speed of 0.9 km/s to the northeast.

After traveling through Mexico, the shadow reaches the American southwest where it enters southern Arizona, New Mexico and western Texas (Figure 4). El Paso lies just south of the center line and will witness a 5 minutes 40 second annular eclipse beginning 16:07 UT. Continuing through the panhandle of Texas, the path enters Oklahoma where Oklahoma City lies just inside the southern limit. Kansas and Missouri are the next two states in the shadow's path. Kansas City lies just outside the northern limit while Springfield barely lies within the southern limit. Both cities witness maximum eclipse at 16:45 UT with the Sun ~62° above the horizon. St. Louis, Springfield and Decatur also lie in the path of annularity, while Chicago and Indianapolis lie just outside the northern and southern limits, respectively (Figure 5).

The instant of greatest eclipse[1] occurs at 17:11:27 UT when the length of annularity reaches its maximum duration of 6 minutes 13 seconds. The Sun's altitude is then 66° and the path is 230 kilometers wide. Toledo stands on the center line as the anti-umbra heads across Lake Erie. Here, observers along both Canadian and U. S. shores of Lake Erie and Lake Ontario will witness the annular phase. Toronto, the largest Canadian city in the path, lies just inside the northern limit where maximum eclipse occurs at 17:24 UT. After skirting northwestern Pennsylvania, the path crosses upper New York State where Buffalo and Rochester witness the annular phase. Most of Vermont and New Hampshire fall within the path limits which continue across the southern half of Maine. Returning to Canada, the shadow crosses southern New Brunswick and Nova Scotia. At 17:56 UT, Halifax experiences an annular phase lasting 5 minutes 53 seconds with a solar altitude of 55°.

As the shadow leaves North America, it sweeps across the North Atlantic where it reaches the Azores at approximately 18:45 UT (Figure 6). The Sun is now 27° high, the path width has grown to 270 kilometers and the central duration has diminished to 5 minutes 10 seconds. Several minutes later, the anti-umbra reaches the Atlantic coast of Morocco and heads inland where the leading edge of the shadow leaves Earth's surface at 18:57 UT. Casablanca witnesses the rare 'ring of fire' as the 4 minute 32 second annular eclipse occurs just 3° above the western horizon. Finally, the annular eclipse ends at 19:02 UT as the trailing edge of the shadow leaves Earth along the sunset terminator. In a period of 3 hours and 42 minutes, the Moon's shadow sweeps along a path almost 14,000 kilometers long, encompassing 0.72 % of Earth's surface area.

---

[1] The instant of greatest eclipse occurs when the distance between the Moon's shadow axis and Earth's geocenter reaches a minimum. Although greatest eclipse differs slightly from the instants of greatest magnitude and greatest duration (for total eclipses), the differences are usually negligible.

GENERAL MAPS OF THE ECLIPSE PATH

ORTHOGRAPHIC PROJECTION MAP OF THE ECLIPSE PATH

Figure 1 is an orthographic projection map of Earth [adapted from Espenak, 1987] showing the path of penumbral (partial) and umbral (annular) eclipse. The daylight terminator is plotted for the instant of greatest eclipse with north towards the top. The sub-Earth point is centered over the point of greatest eclipse and is marked by a '*' or asterisk. Earth's sub-solar point at that instant is indicated by a '✿' or star shaped character.

The limits of the Moon's penumbral shadow delineate the region of visibility of the partial solar eclipse. This irregular or saddle shaped region often covers more than half of the daylight hemisphere of Earth and consists of several distinct zones or limits. At the northern and/or southern boundaries lie the limits of the penumbra's path. Partial eclipses have only one of these limits, as do central eclipses when the shadow axis falls no closer than about 0.45 radii from Earth's center. Great loops at the western and eastern extremes of the penumbra's path identify the areas where the eclipse begins/ends at sunrise and sunset, respectively. If the penumbra has both a northern and southern limit, the rising and setting curves form two separate, closed loops. Otherwise, the curves are connected in a distorted figure eight. Bisecting the 'eclipse begins/ends at sunrise and sunset' loops is the curve of maximum eclipse at sunrise (western loop) and sunset (eastern loop). The points **P1** and **P4** mark the coordinates where the penumbral shadow first contacts (partial eclipse begins) and last contacts (partial eclipse ends) Earth's surface. If the penumbral path has both a northern and southern limit (as does the May 1994 eclipse), then points **P2** and **P3** are also plotted. These correspond to the coordinates where the penumbral shadow cone becomes internally tangent to Earth's disk.

A curve of maximum eclipse is the locus of all points where the eclipse is at maximum at a given time. Curves of maximum eclipse are plotted at each half hour Universal Time. They generally run from the northern to the southern penumbral limits, or from the maximum eclipse at sunrise and sunset curves to one of the limits. The curves of maximum eclipse run through the half-hourly outlines of the umbral shadow, from which the Universal Time of each curve can be identified. The curves of constant eclipse magnitude[2] delineate the locus of all points where the magnitude at maximum eclipse is constant. These curves run exclusively between the curves of maximum eclipse at sunrise and sunset. Furthermore, they're parallel to the northern/southern penumbral limits and the umbral paths of central eclipses. The northern and southern limits of the penumbra may be thought of as curves of constant magnitude of 0.0. The adjacent curves are for magnitudes of 0.2, 0.4, 0.6 and 0.8. For total eclipses, the northern and southern limits of the umbra are curves of constant magnitude of 1.0. Umbral path limits for annular eclipses are curves of maximum eclipse magnitude. The magnitude is always less than 1.0 for annular eclipses.

In the upper left corner of Figure 1 are the Universal Times of greatest eclipse and conjunction of the Moon and Sun in right ascension, followed by the minimum distance of the Moon's shadow axis from Earth's center in Earth radii **GAMMA** and the geocentric ratio of diameters of the Moon and the Sun **RATIO**. To the upper right are exterior contact times of the Moon's shadow with Earth. **P1** and **P4** are the first and last contacts of the penumbra; they mark the start and end of the partial eclipse. **U1** and **U4** are the first and last contacts of the umbra; they denote the start and end of the annular eclipse. Below the map are the geocentric coordinates of the Sun and Moon at the instant of greatest eclipse. They include of the right ascension **RA**, declination **DEC**, apparent semi-diameter **SD** and equatorial horizontal parallax **HP**. The Saros series for the eclipse is listed, followed by a pair of numbers in parentheses. The first number identifies the sequence order of the eclipse in the Saros, while the second is the total number of eclipses in the series. The Julian Date **JD** at greatest eclipse is given, followed by the extrapolated value of ΔT[3] used in the calculations. Finally, the geodetic coordinates of the point of greatest eclipse are given, as well as the local circumstances there. In particular, the Sun's altitude **ALT** and azimuth **AZ** are listed along with the duration of umbral eclipse (minutes:seconds) and the width of the path (kilometers).

---

[2] Eclipse magnitude is defined as the fraction or percentage of the Sun's diameter occulted by the Moon. It's usually expressed at greatest eclipse. Eclipse magnitude is strictly a ratio of *diameters* and should not be confused with eclipse obscuration which is a measure of the Sun's surface *area* occulted by the Moon.

[3] ΔT is the difference between Terrestrial Dynamical Time and Universal Time

STEREOGRAPHIC PROJECTION MAP OF THE ECLIPSE PATH

The stereographic projection of Earth in Figure 2 depicts the path of penumbral and umbral eclipse in greater detail. The map is oriented with the point of greatest eclipse near the center and north is towards the top. International political borders are shown and circles of latitude and longitude at plotted at 20° increments. The saddle shaped region of penumbral or partial eclipse includes labels identifying the northern and southern limits, curves of eclipse begins or ends at sunrise, curves of eclipse begins or ends at sunset, and curves of maximum eclipse at sunrise and sunset. Curves of constant eclipse magnitude are plotted for 20%, 40%, 60% and 80%, as are the limits of the annular path. Also included are curves of greatest eclipse for every thirty minutes Universal Time.

Figure 2 may be used to quickly determine the approximate time and magnitude of greatest eclipse for any location from which the eclipse is visible.

EQUIDISTANT CONIC PROJECTION MAPS OF THE ECLIPSE PATH

Figures 3, 4, 5 and 6 are equidistant conic projection maps which isolate specific regions of the eclipse path. The projection was selected to minimize distortion over the regions depicted. Once again, curves of maximum eclipse and constant eclipse magnitude are plotted along with identifying labels. A linear scale is included for estimating approximate distances (kilometers) in each figure. Within the northern and southern limits of the annular path, the outline of the umbral shadow is plotted at ten minute intervals. Figures 4, 5 and 6 are drawn at the same scale (~1:12,270,000) and include the center line as well as the positions of many of the larger cities or metropolitan areas in and near the central path. The size of each city is logarithmically proportional to its population.

ELEMENTS, SHADOW CONTACTS AND ECLIPSE PATH TABLES

The geocentric ephemeris for the Moon and Sun, various parameters and constants used in the predictions, the besselian elements (polynomial form) are given in Table 1. The eclipse predictions and elements were derived from solar and lunar data contained in the DE200 and LE200 ephemerides developed jointly by the Jet Propulsion Laboratory and the U. S. Naval Observatory for use in the *Astronomical Almanac* for 1984 and after. Unless otherwise stated, all predictions are based on center of mass positions for the Sun and Moon with no corrections made for center of figure, center of motion, lunar limb profile or atmospheric refraction. Furthermore, these predictions depart from IAU convention by using a smaller constant for the mean lunar radius k for all umbral contacts (see: LUNAR LIMB PROFILE). Times are expressed in either Terrestrial Dynamical Time (TDT) or in Universal Time (UT) where the best value of ΔT available at the time of preparation is used.

Table 2 lists all external and internal contacts of penumbral and umbral shadows with Earth. They include TDT times and geodetic coordinates with and without corrections for ΔT. The external contacts of the penumbral **P1** and **P4** mark the instants when the partial eclipse begins and ends, respectively. The external contacts of the umbral **U1** and **U4** mark the instants when the umbral eclipse begins and ends. Likewise, the extremes of the penumbral and umbral paths, and extreme limits of the center line are given. The IAU longitude convention is used throughout this publication (i.e. - eastern longitudes are positive; western longitudes are negative; negative latitudes are south of the Equator).

The path of the umbral shadow is delineated at five minute intervals of Universal Time in Table 3. The coordinates of the northern limit, the southern limit and the center line are listed to the nearest tenth of an arc-minute (~185 m at the Equator). The Sun's altitude and azimuth, the path width and umbral duration are calculated for the center line coordinates. Table 4 presents a physical ephemeris for the umbral shadow at five minute intervals of Universal Time. The center line coordinates are followed by the topocentric ratio of the apparent diameters of the Moon and Sun, the eclipse obscuration[4], and the Sun's altitude at that instant. The path azimuth differs from the Sun's azimuth and represents the direction of the umbral shadow's motion projected onto the surface of the Earth. The central path width, the umbral shadow's major and minor axes and its instantaneous velocity with respect to Earth's surface are included. Finally, the center line duration of the annular phase is given.

Local circumstances for each center line position listed in Tables 3 and 4 are presented in Table 5. The first three columns give the Universal Time of maximum eclipse, the center line duration of annularity and the altitude of the Sun at that instant. The following columns list each of the four eclipse contact

---

[4] Eclipse obscuration is defined as the fraction of the Sun's surface area occulted by the Moon.

3

times followed by their related contact position angles and the corresponding altitude of the Sun. The four contacts[5] identify significant stages in the progress of the eclipse. The position angles **P** and **V** identify the point along the Sun's disk where each contact occurs[6]. The altitude of the Sun at second and third contact is omitted since it's always within 1° of the altitude at maximum eclipse (column 3).

Table 6 presents topocentric values at maximum eclipse for the Moon's horizontal parallax, semidiameter, relative angular velocity with respect to the Sun, and libration in longitude. The altitude and azimuth of the Sun are given along with the azimuth of the umbral path. The northern limit position angle identifies the point on the lunar disk defining the umbral path's northern limit. It's measured counterclockwise from the north point of the lunar disk. In addition, corrections to the path limits due to the lunar limb profile are listed. The irregular profile of the Moon results in a zone of 'grazing eclipse' at each limit which is delineated by interior and exterior contacts of lunar features with the Sun's limb. The section LIMB CORRECTIONS TO THE PATH LIMITS: GRAZE ZONES describes this geometry in greater detail. Corrections to the center line durations due to the lunar limb profile are also included. When added to the durations in Tables 3, 4, 5 and 7, a slightly shorter central annular phase is predicted.

To aid and assist in the plotting of the umbral path on large scale maps, the path coordinates are also tabulated at 1° intervals in longitude in Table 7. The latitude of the northern limit, southern limit and center line for each longitude is tabulated along with the Universal Time of maximum eclipse at each position. Finally, local circumstances on the center line at maximum eclipse are listed and include the Sun's altitude and azimuth, the umbral path width and the central duration of annularity.

LOCAL CIRCUMSTANCES TABLES

Local circumstances from over 900 cities, metropolitan areas and places in North America, Europe and Africa are presented in Tables 9 through 14. Each table is broken down into two parts. The first part, labeled **a**, appears on even numbered pages and gives circumstances at maximum eclipse[7] for each location. The coordinates are listed along with the location's elevation (meters) above sea-level, if known. If the elevation is unknown (i.e. - not in the data base), then the local circumstances for that location are calculated at sea-level. In any case, the elevation does not play a significant role in the predictions unless the location is near the umbral path limits and the Sun's altitude is relatively small (>20°). The Universal Time of maximum eclipse (either partial or annular) is listed to an accuracy of 0.1 seconds. If the eclipse is annular, then the umbral duration and the path width are given. Next, the altitude and azimuth of the Sun at maximum eclipse are listed along with the position angles **P** and **V** of the Moon's disk with respect to the Sun. Finally, the magnitude and obscuration are listed at the instant of maximum eclipse. Note that for umbral eclipses (annular and total), the eclipse magnitude is identical to the topocentric ratio of the Moon's and Sun's apparent diameters. Furthermore, the eclipse magnitude is always less than 1 for annular eclipses and equal to or greater than 1 for total eclipses.

The second part of each table, labeled **b**, is found on odd numbered pages. It gives local circumstances for each location listed on the facing page at each contact during the eclipse. The Universal Time of each contact is given along with the altitude of the Sun, followed by position angles **P** and **V**. These angles identify the point along the Sun's disk where each contact occurs and are measured counterclockwise from the north and zenith points, respectively. Locations outside the umbral path miss the umbral eclipse and only witness first and fourth contacts. The effects of refraction have included in these calculations although no correction has been applied for center of figure or the lunar limb profile.

Locations were chosen based on position near the central path, general geographic distribution and population. The primary source for geographic coordinates is *The New International Atlas* (Rand McNally, 1991). Elevations for major cites were taken from *Climates of the World* (U. S. Dept. of Commerce, 1972). In this rapidly changing political world, it is often difficult to ascertain the correct name

---

[5] First contact is defined as the instant of external tangency between the Sun and Moon; it marks the beginning of the partial eclipse.

Second and third contacts define the two instants of internal tangency between the Sun and Moon; they signify the commencement and termination of the umbral (total or annular) phase.

Fourth contact is the instant of last external contact and it marks the end of the partial eclipse.

[6] P is defined as the contact angle measured counter-clockwise from the *north* point of the Sun's disk.

V is defined as the contact angle measured counter-clockwise from the *zenith* point of the Sun's disk.

[7] For partial eclipses, maximum eclipse is the instant when the greatest fraction of the Sun's diameter is occulted. For umbral eclipses (total or annular), maximum eclipse is the instant of mid-totality or mid-annularity.

or spelling for a given location. Therefore, the information presented here is for location purposes only and is not meant to be authoritative. Furthermore, it does not imply recognition of status of any location by the United States Government. Corrections to names, spellings, coordinates and elevations is solicited in order to update the geographic data base for future eclipse predictions.

DETAILED MAPS OF THE UMBRAL PATH

The path of annularity has been plotted by hand on a set of eight detailed maps appearing in the last section of this publication. The maps are Global Navigation and Planning Charts or GNC's from the Defense Mapping Agency which use a Lambert conformal conic projection. More specifically, GNC-2 covers the North American section of the path while GNC-11 covers Africa. GNC's have a scale of 1:5,000,000 (1 inch ~ 69 nautical miles), which is adequate for showing major cities, highways, airports, rivers, bodies of water and basic topography required for eclipse expedition planning including site selection, transportation logistics and weather contingency strategies.

Northern and southern limits as well as the center line of the path are draw using predictions from Table 3. No corrections have been made for center of figure or lunar limb profile. However, such corrections have little or no effect at this scale. Although, atmospheric refraction has not been included, it's effects play a significant role only at low solar altitudes (i.e. - Morocco). In any case, refraction corrections to the path are uncertain since they depend on the atmospheric temperature-pressure profile which cannot be predicted in advance. If observations from the graze zones are planned, then the path must be plotted on higher scale maps using limb corrections in Table 6. See PLOTTING THE PATH ON MAPS for sources and more information. The GNC paths also depict the curve of maximum eclipse at five minute increments in Universal Time [Table 3].

ESTIMATING TIMES OF SECOND AND THIRD CONTACTS

The times of second and third contact for any location not listed in this publication can be estimated using the detailed maps found in the final section. Alternatively, the contact times can be estimated from maps on which the umbral path has been plotted. Table 7 lists the path coordinates conveniently arranged in 1° increments of longitude to assist plotting by hand. The path coordinates in Table 3 define a line of maximum eclipse at five minute increments in time. These lines of maximum eclipse each represent the projection diameter of the umbral shadow at the given time. Thus, any point on one of these lines will witness maximum eclipse (i.e.: mid-annularity) at the same instant. The coordinates in Table 3 should be added to the map in order to construct lines of maximum eclipse.

The estimation of contact times for any one point begins with an interpolation for the time of maximum eclipse at that location. The time of maximum eclipse is proportional to a point's distance between two adjacent lines of maximum eclipse, measured along a line parallel to the center line. This relationship is valid along most of the path with the exception of the extreme ends where the shadow experiences its largest acceleration. The center line duration of annularity D and the path width W are similarly interpolated from the values of the adjacent lines of maximum eclipse as listed in Table 3. Since the location of interest probably does not lie on the center line, it's useful to have an expression for calculating the duration of annularity d as a function of its perpendicular distance a from the center line:

$$d = D (1 - (2 a/W)^2)^{1/2} \text{ seconds} \qquad [1]$$

where: D = duration of annularity on the center line (seconds)
W = width of the path (kilometers)
a = perpendicular distance from the center line (kilometers)

If t_m is the interpolated time of maximum eclipse for the location, then the approximate times of second and third contacts (t_2 and t_3, respectively) are:

Second Contact:	$t_2 = t_m - d/2$	[2]
Third Contact:	$t_3 = t_m + d/2$	[3]

The position angles of second and third contact (either P or V) for any location off the center line are also useful in some applications. First, linearly interpolate the center line position angles of second and

5

third contacts from the values of the adjacent lines of maximum eclipse as listed in Table 5. If X_2 and X_3 are the interpolated center line position angles of second and third contacts, then the position angles x_2 and x_3 of those contacts for an observer located a kilometers from the center line are:

Second Contact:	$x_2 = X_2 - ArcSin\ (2\ a/W)$	[4]
Third Contact:	$x_3 = X_3 + ArcSin\ (2\ a/W)$	[5]

where: X_n = the interpolated position angle (either P or V) of contact n on center line

x_n = the interpolated position angle (either P or V) of contact n at location

D = duration of annularity on the center line (seconds)

W = width of the path (kilometers)

a = perpendicular distance from the center line (kilometers)

(use negative values for locations south of the center line)

MEAN LUNAR RADIUS

A fundamental parameter used in the prediction of solar eclipses is the Moon's mean radius k, expressed in units of Earth's equatorial radius. The actual radius of the Moon varies as a function of position angle and libration due to the irregularity of the lunar limb profile. From 1968 through 1980, the Nautical Almanac Office used two separate values for k in their eclipse predictions. The larger value (k=0.2724880) representing a mean over lunar topographic features was used for all penumbral (i.e. - exterior) contacts and for annular eclipses. A smaller value (k=0.272281) representing a mean minimum radius was reserved exclusively for umbral (i.e. - interior) contact calculations of total eclipses [*Explanatory Supplement*, 1974]. Unfortunately, the use of two different values of k for umbral eclipses introduces a discontinuity in the case of hybrid or annular-total eclipses.

In August 1982, the IAU General Assembly adopted a value of k=0.2725076 for the mean lunar radius. This value is currently used by the Nautical Almanac Office for all solar eclipse predictions [Fiala and Lukac, 1983] and is believed to be the best mean radius, averaging mountain peaks and low valleys along the Moon's rugged limb. In general, the adoption of one single value for k is commendable because it eliminates the discontinuity in the case of annular-total eclipses and ends confusion arising from the use of two different values. However, the use of even the best 'mean' value for the Moon's radius introduces a problem in predicting the character and duration of umbral eclipses, particularly total eclipses. A total eclipse can be defined as an eclipse in which the Sun's disk is completely occulted by the Moon. This cannot occur so long as any photospheric rays are visible through deep valleys along the Moon's limb [Meeus, Grosjean and Vanderleen, 1966]. But the use of the IAU's mean k guarantees that some annular or annular-total eclipses will be misidentified as total. A case in point is the eclipse of 3 October 1986. The *Astronomical Almanac* identified this event as a total eclipse of 3 seconds duration when in it was in fact a beaded annular eclipse. Clearly, a smaller value of k is needed since it is more representative of the deepest lunar valley floors, hence the minimum solid disk radius and ensures that an eclipse is truly total.

Of primary interest to most observers are the times when central eclipse begins and ends (second and third contacts, respectively) and the duration of the central phase. When the IAU's mean value for k is used to calculate these times, they must be corrected to accommodate low valleys (total) or high mountains (annular) along the Moon's limb. The calculation of these corrections is not trivial but must be performed, especially if one plans to observe near the path limits [Herald, 1983]. For observers near the center line of a total eclipse, the limb corrections can be closely approximated by using a smaller value of k which accounts for the valleys along the profile.

This work uses the IAU's accepted value of k (k=0.2725076) for all penumbral (exterior) contacts. In order to avoid eclipse type misidentification and to predict central durations which are closer to the actual durations observed at total eclipses, we depart from convention by adopting the smaller value for k (k=0.272281) for all umbral (interior) contacts. This is consistent with predictions published in *Fifty Year Canon of Solar Eclipses: 1986 - 2035* [Espenak, 1987]. Consequently, the smaller k produces shorter umbral durations and narrower paths for total eclipses when compared with calculations using the IAU value for k. Similarly, the smaller k predicts longer umbral durations and wider paths for annular eclipses.

LUNAR LIMB PROFILE

Eclipse contact times, the magnitude and the duration of annularity all ultimately depend on the angular diameters and relative velocities of the Sun and the Moon. Unfortunately, these calculations are limited in accuracy by the departure of the Moon's limb from a perfectly circular figure. The Moon's surface exhibits a rather dramatic topography which manifests itself as an irregular limb when seen in profile. Most eclipse calculations assume some mean lunar radius which averages high mountain peaks and low valleys along the Moon's rugged limb. Such an approximation is acceptable for many applications, but if higher accuracy is needed, the Moon's actual limb profile must be considered. Fortunately, an extensive body of knowledge exists on this subject in the form of Watt's limb charts [Watts, 1963]. These data are the product of a photographic survey of the marginal zone of the Moon and give limb profile heights with respect to an adopted smooth reference surface (or datum). Analyses of lunar occultations of stars by Van Flandern [1970] and Morrison [1979] have shown that the average cross-section of Watts' datum is slightly elliptical rather than circular. Furthermore, the implicit center of the datum (i.e. - the center of figure) is displaced from the Moon's center of mass. In a follow-up analysis of 66000 occultations, Morrison and Appleby [1981] have found that the radius of the datum appears to vary with libration. These variations produce systematic errors in Watts' original limb profile heights which attain 0.4 arc-seconds at some position angles. Thus, corrections to Watts' limb profile data are necessary to ensure that the reference datum is a sphere with its center at the center of mass.

The Watts charts have been digitized by Her Majesty's Nautical Almanac Office in Herstmonceux, England, and transformed to grid-profile format at the U. S. Naval Observatory. In this computer readable form, the Watts limb charts lend themselves to the generation of limb profiles for any lunar libration. Ellipticity and libration corrections may be applied to refer the profile to the Moon's center of mass. Such a profile can then be used to correct eclipse predictions which have been generated using a mean lunar limb.

Along the eclipse path, the Moon's topocentric libration (physical + optical libration) in longitude ranges from l=-0.7° to l=-2.5°. Thus, a limb profile with the appropriate libration is required in any detailed analysis of contact times, central duration's, etc.. Nevertheless, a profile with an intermediate libration is valuable for general planning for any point along the path. The center of mass corrected lunar limb profile presented in Figure 7 is for the center line at the instant of greatest eclipse (17:11:27 UT). At that time, the Moon's topocentric librations are l=-1.65°, b=-0.12° and c=-17.18°, and the apparent topocentric semi-diameters of the Sun and Moon are 950.3 and 896.2 arc-seconds respectively. The Moon's angular velocity is 0.289 arc-seconds per second with respect to the Sun.

The radial scale of the profile in Figure 7 (see scale to upper left) is greatly exaggerated so that the true limb's departure from the mean lunar limb is readily apparent. The mean limb with respect to the center of figure of Watts' original data is shown along with the mean limb with respect to the center of mass. Note that all the predictions presented in this paper are calculated with respect to the latter limb unless otherwise noted. Position angles of various lunar features can be read using the protractor in the center of the diagram. The position angles of second and third contact are clearly marked along with the north pole of the Moon's axis of rotation and the observer's zenith at mid-annularity. The dashed line arrows identify the points on the limb which define the northern and southern limits of the path. To the upper left of the profile are the Moon's mean lunar radius k (expressed in Earth equatorial radii), topocentric semi-diameter SD and horizontal parallax HP. As discussed in the section MEAN LUNAR RADIUS, the Moon's mean radius k (k=0.2722810) is smaller than the adopted IAU value (k=0.2725076). To the upper right of the profile are the Sun's semi-diameter SUN SD, the angular velocity of the Moon with respect to the Sun VELOC. and the position angle of the path's northern/southern limit axis LIMITS. In the lower right are the Universal Times of the four contacts and maximum eclipse. The geographic coordinates and local circumstances at maximum eclipse are given along the bottom of the figure.

In investigations where accurate contact times are needed, the lunar limb profile can be used to correct the nominal or mean limb predictions. For any given position angle, there will be a high mountain (annular eclipses) or a low valley (total eclipses) in the vicinity which ultimately determines the true instant of contact. The difference, in time, between the Sun's position when tangent to the contact point on the mean limb and tangent to the highest mountain (annular) or lowest valley (total) at actual contact is the desired correction to the predicted contact time. On the exaggerated radial scale of Figure 7, the Sun's limb can be represented as an epicyclic curve which is tangential to the mean lunar limb at the point of contact and departs from the limb by **h** as follows:

$$\mathbf{h} = \mathbf{S} \ (\mathbf{m}\text{-}1) \ (1\text{-}\cos[\mathbf{C}]) \qquad\qquad [6]$$

where: S = the Sun's semi-diameter
m = the eclipse magnitude
C = the angle from the point of contact

Herald [1983] has taken advantage of this geometry to develop a graphical procedure for estimating correction times over a range of position angles. Briefly, a displacement curve of the Sun's limb is constructed on a transparent overlay by way of equation [6]. For a given position angle, the solar limb overlay is moved radially from the mean lunar limb contact point until it is tangent to the lowest lunar profile feature in the vicinity. The solar limb's distance **d** (arc-seconds) from the mean lunar limb is then converted to a time correction Δ by:

$$\Delta = \mathbf{d} \ \mathbf{v} \ \cos[\mathbf{X} - \mathbf{C}] \qquad\qquad [7]$$

where: d = the distance of Solar limb from mean lunar limb (arc-sec)
v = the angular velocity of the Moon with respect to the Sun (arc-sec/sec)
X = the center line position angle of the contact
C = the angle from the point of contact

This operation may be used for predicting the formation and location of Bailey's beads. When calculations are performed over large range of position angles, a contact time correction curve can then be constructed.

Since the limb profile data are available in digital form, an analytic solution to the problem is possible which is straight forward and quite robust. Curves of corrections to the times of second and third contact for most position angles have been computer generated and are plotted in Figure 7. In interpreting these curves, the circumference of the central protractor functions as the nominal or mean contact time (using the Moon's mean limb) as a function of position angle. The departure of the correction curve from the mean contact time can then be read directly from Figure 7 for any position angle by using the radial scale in the upper right corner (units in seconds of time). Time corrections external to the protractor (most second contact corrections) are added to the mean contact time; time corrections internal to the protractor (all third contact corrections) are subtracted from the mean contact time.

Across most of North America, the Moon's topocentric libration in longitude at maximum eclipse is within half a degree of its value at greatest eclipse. Therefore, the limb profile and contact correction time curves in Figure 7 may be used in all but the most critical investigations.

LIMB CORRECTIONS TO THE PATH LIMITS: GRAZE ZONES

The northern and southern umbral limits provided in this publication were derived using the Moon's center of mass and a mean lunar radius. They have not been corrected for the Moon's center of figure or the effects of the lunar limb profile. In applications where precise limits are required, Watt's limb data must be used to correct the nominal or mean path. Unfortunately, a single correction at each limit is not possible since the Moon's libration in longitude and the contact points of the limits along the Moon's limb each vary as a function of time and position along the umbral path. This makes it necessary to calculate a unique correction to the limits at each point along the path. Furthermore, the northern and southern limits of the umbral path are actually paralleled by a relatively narrow zone where the eclipse is neither penumbral nor umbral. An observer positioned here will witness a solar crescent which is fragmented into a series of bright beads and short segments whose morphology changes quickly with the rapidly varying geometry of the Moon with respect to the Sun. These beading phenomena are caused by the appearance of photospheric rays which alternately pass through deep lunar valleys and hide behind high mountain peaks as the Moon's irregular limb grazes the edge of the Sun's disk. The geometry is directly analogous to the case of grazing occultations of stars by the Moon. The graze zone is typically five to ten kilometers wide and its interior and exterior boundaries can be predicted using the lunar limb profile. The interior boundaries define the actual limits of the umbral eclipse (both total and annular) while the exterior boundaries set the outer limits of the grazing eclipse zone.

Table 6 provides topocentric data and corrections to the path limits due to the true lunar limb profile. At five minute intervals, the table lists the Moon's topocentric horizontal parallax, the semi-diameter, the relative angular velocity of the Moon with respect to the Sun and lunar libration in longitude. The center line altitude and azimuth of the Sun is given, followed by the azimuth of the umbral path. The

8

position angle of the point on the Moon's limb which defines the northern limit of the path is measured counter-clockwise (i.e. - eastward) from the north point on the limb. The path corrections to the northern and southern limits are listed as interior and exterior components in order to define the graze zone. Positive corrections are in the northern sense while negative shifts are in the southern sense. These corrections [minutes of arc in latitude] may be added directly to the path coordinates listed in Table 3. Corrections to the center line umbral durations due to the lunar limb profile are also included and they are all negative. Thus, when added to the central durations given in Tables 3, 4, 5 and 7, a slightly shorter central annular phase is predicted.

SAROS HISTORY

The annular eclipse of 10 May 1994 is the fifty-seventh member of Saros series 128, as defined by van den Bergh (1955). All eclipses in the series occur at the Moon's descending node and gamma[8] increases with each member in the series. The family began on 29 Aug 984 with a partial eclipse in the southern hemisphere. During the next four centuries, a total of twenty-four partial eclipses occurred with the eclipse magnitude of each succeeding event gradually increasing. Finally, the first umbral eclipse occurred on 16 May 1417. The event was a total eclipse of short duration which was followed by three more eclipses with similar characteristics. The twenty-ninth event occurred on 28 Jun 1489 and was of annular/total nature. The ensuing three members were also annular/total eclipses of monotonically decreasing duration. This was a direct consequence of the Moon's increasing distance with each event. Eventually, the character of the series changed to pure annular with the thirty-third member on 11 Aug 1561.

For almost three centuries, Saros 128 continued to produce annular eclipses where each event was of progressively smaller magnitude and increasing duration. The trend culminated with the annular eclipses of 1 Feb 1832 and 12 Feb 1850, each of which was characterized by an eclipse magnitude of 0.934 and a greatest duration of 8m 35s. Having reached its orbital apogee, the Moon's distance began to decrease with each succeeding eclipse, resulting in annular eclipses of increasing magnitude and decreasing duration. Although the annular eclipse of 10 May 1994 still reaches a magnitude of 0.943 and a maximum duration of 6 minutes 14 seconds, each subsequent event is of rapidly diminishing maximum duration. To illustrate, the eclipses of 20 May 2012, 1 Jun 2030 and 11 Jun 2048 will exhibit maximum durations of 5m 46s, 5m 21s and 4m 58s respectively as the path of each event shifts northward. Member sixty-four is the last central eclipse of the series and occurs on 15 Jul 2120. The remaining nine events are partial in the northern hemisphere with the last and seventy-third member occurring on 1 Nov 2282.

In summary, Saros series 128 includes 73 eclipses with the following distribution:

	Partial	*Annular*	*Ann/Total*	*Total*
Non-Central	33	0	0	0
Central	—	32	4	4

---

[8] Gamma is measured in Earth radii and is the minimum distance of the Moon's shadow axis from Earth's center during an eclipse. This occurs at and defines the instant of greatest eclipse. Gamma takes on negative values when the shadow axis is south of the Earth's center.

WEATHER PROSPECTS FOR THE ECLIPSE

OVERVIEW

The mid latitude track of the May 1994 eclipse takes it across some of the more active weather regions of the Northern Hemisphere. At its southernmost extent over Mexico, weather patterns are regular and reliable, with relatively small variation from season to season. At the apex of its path over Nova Scotia, weather is cloudy and changeable. And at sunset over Morocco, there is a combination of both - the sunny disposition of the sub tropics combined with the vagaries of passing temperate zone lows and highs.

MEXICO

The eclipse track crosses one of the driest and sunniest parts of Mexico, in spite of a jumbled terrain which includes cool beaches, deserts, 3000 meter peaks and a broad 1500 meter plateau. Each of these features has an influence on the weather, but the moisture supply is so low in most areas that, with only one exception, cloud cover is sparse and sunshine plentiful. The main control on the weather is a large and permanent high pressure cell which resides in the eastern Pacific about half way between Hawaii and San Francisco. This semi-permanent anticyclone suppresses rain bearing weather systems along the California and Baja coasts, bringing the dry summers for which the area is well known.

During the winter months, weather systems arrive over the eclipse track on upper level westerly winds. During the summer, easterly low level trade winds carry the moisture which builds thunderstorms and brings the rainy season or "tiempo des aguas". May is the intermediate season, too soon for the moist easterlies, and getting very late for westerly disturbances. Still, what weather occurs on eclipse day is likely to come with the last westerly troughs moving in from the Pacific.

WESTERN BAJA

Although its one of the drier places in the world, the west coast of the Baja Peninsula is plagued by a persistent low level cloudiness and fog. The dull grey skies, mostly in the morning, come courtesy of the high pressure system which camps in the eastern Pacific. Wind circulations around the high build a strong temperature inversion which traps moisture in the lowest 2000 meters of the atmosphere.

A cold California Current flows southward along the coast, lowering the air temperature, and bringing the atmosphere to saturation. Winds, mostly out of northwest, carry the cloud onshore to plague eclipse chasers (as many discovered in July 1991). The sun is usually able to burn this cloud away by noon, but that will be much too late for this eclipse. Depending on the lie of the coast, some areas are more prone than others to intercept the flow of moist air from the ocean. First among these is the prominent fishhook peninsula jutting from the Baja coast, terminating at Punta Eugenia. The hook creates a large bay - Bahia Sebastian Vizcaino - which scoops the moist northwest flow from the ocean and directs it inland. And it's here that the eclipse first comes ashore.

Satellite pictures show that low cloud and fog often fill the entire Desierto de Vizcaino which lies at the bottom of the Bay. Statistics collected over a 30 year period show that the cloud does not usually penetrate as far as San Ignacio but it doesn't miss by much. Nearby El Alamo records fog nearly 6 days of the month. Some of the cloudiest areas along the Baja coast report fog on one day out of three. If there were a station on the coast of the Bahia Sebastian Vizcaino it is likely that it would be a strong challenger for the foggiest location.

Those who wish to challenge the statistics and be the first to greet the ringed sun on the west side of the Baja should keep a weather eye peeled for evidence of impending cloudiness. While fog and low cloud are obvious, even a feeling of humidity or mugginess may signal that cloud will form once the sun rises. Be wary of cloud and fog in the distance, for the rising sun will quickly warm the ground and strengthen the onshore breezes which will carry the cloud inland.

All-in-all, eclipse chasers would do well to avoid much of western Baja, and seek better weather prospects elsewhere. And better prospects are only a short distance away, on the east slopes of the peninsula and along its mountain spine.

ALONG THE GULF OF CALIFORNIA

The sunniest skies along the entire eclipse track can be found on the eastern slopes of the Baja Peninsula and the western coast of mainland Mexico. Figure 8 shows that the frequency of clear skies (less than 1/3 cloud cover) in May ranges from 80 to 90 percent[9]. The Gulf of California is protected in the east by the Sierra Madre Occidental and in the west by the mountain backbone of the Baja Peninsula. Winds arriving from nearly every direction have to travel downslope - drying and losing what little cloud might remain. Only high level disturbances can cross the mountains undisturbed and the climatological record shows that most of the cloudiness comes from ice crystal cirrus clouds. These clouds tend to be thinner than their water vapor cousins. The eclipse should be visible through cirrus cloud, and the cloud might even add an attractive element to the sky.

Roads along the eastern side of the Baja Peninsula are limited and mostly unpaved. Fortunately it's only necessary to travel just far enough inland to get away from the Pacific marine cloud. El Arco, just south of the center line and on the main highway is probably blessed with good prospects. For those who want to capture Bailey's beads from the north limit, the skies around Punta Prieta are very promising.

On the east side of the Gulf of California access is almost unlimited and the weather is excellent. Hermosillo and Guaymas show clear skies or scattered cloud on 77% of the days in May. Prospects are even better along the beaches of the Gulf, with the coastal pueblo of Bahia Kino likely offering the best weather prospects of any position along the entire eclipse track. It's also very close to the center line.

MAINLAND MEXICO

Climbing the Sierra Madre Occidental, the eclipse track moves onto the Interior Plateau of mainland Mexico. The height of the Sierra, 3200 meters high, keeps all but the largest Pacific systems from bringing precipitation. These Pacific disturbances, borne on westerly winds in the upper atmosphere, are primarily a winter feature, and in May are becoming increasingly rare. Cloud cover increases slowly as the track travels from Hermosillo to El Paso in Texas. Some of this increase in cloud cover is a result of heating of the eastward facing mountain slopes. Bubbles of warm air cool as they rise, becoming saturated and turning into puffy cumulus. At worst, only scattered cumulus can be expected since eclipse time is around 10:15 AM CST in northern Mexico. Cloud development doesn't peak until 2 PM when temperatures are approaching their maximum.

Morning fog and stratus is more likely to be a problem in the jumble of mountainous terrain. High and dry elevations cool rapidly overnight. Cool air collects in the valleys where temperatures may fall low enough to allow the air to become saturated. Overnight lows average 59⁻F at Hermosillo in May. In comparison to the frequency of foggy mornings along the outer Baja coast, it's a minor problem in the mountains, and can be avoided by moving a short distance. Most overnight fogs will have likely burned off by eclipse time. Fog may redevelop as the ground cools during the eclipse, but the temperature decline in an annular eclipse will be smaller than that during a total. The outcome will depend on the moisture available.

Whether in dry or wet climates, mountains generate cloudiness. The Interior Plateau is no exception. Satellite images from May of 1992 show that the Sierra Madres are often spotted with patchy morning clouds, though usually not too heavily. These clouds are often at middle and high levels - 6000 feet or more above ground. Flatter areas away from the mountains (i.e. - up toward Ciudad Juarez and El Paso, Texas) are probably better sites, but for the most part the cloud which forms over the Sierra will blow toward them anyway. The effects of this cloud can be seen in Figure 8 as the area within the 50% contour which surrounds Temosachic.

During the spring months the sub-tropical jet stream can usually be found over northern Mexico, arcing across the Baja Peninsula and into the southern United States. The jet stream is occasionally marked by a band of cirrus and altostratus cloud, sometimes thick, but usually thin and wispy. Depending on the weather patterns of the day, the jet may lie south of the eclipse track, or a little to the north, though the former is more likely. In general, it's found close to the track, and may be difficult or impossible to avoid.

---

[9] This figure is derived from data collected at selected intervals during the daylight hours. The chart is not directly comparable with Figures 9 to 11 because of national differences in data collection. However the patterns of heavy and light cloudiness are similar.

11

ACROSS THE UNITED STATES AND CANADA

As the eclipse track moves northeastward, it encounters higher latitude weather systems where cloudiness increases steadily. Figures 10 and 11 show the frequency of clear skies (less than 1/3 cloud cover) in May for the U. S. and Canada. Excellent weather greets the umbra as it enters the U. S.. Discouraging statistics bid it farewell over the Atlantic, the Azores and Africa. Even under conditions of scattered cloudiness (<50% sky coverage), the annular eclipse should be visible. Based on climatology only, New Mexico, southeast Arizona and El Paso are the best destinations. May 1992 satellite photos show that the El Paso area had favorable conditions at eclipse time on 21 days. Areas in Mexico faired even better.

The Gulf of Mexico supplies most of the moisture which fuels the weather systems of spring. In May, low level winds have turned to the south across much of the Gulf coast and Great Plains. Warm humid air floods northward to mark the beginning of summer. Thunderstorms often erupt across the States as weather disturbances and frontal systems collide with the humid air. The westward flow of Gulf air is blocked by the Rocky Mountain chain, and is re-directed into Oklahoma, Kansas, Missouri, and eastward to envelop the lower Great Lakes. The Appalachians block the flow to some extent, but the Atlantic Ocean, another moisture source, is ready to fill in when winds turn easterly.

The western edge of this giant atmospheric river lies very nearly along the eclipse track. Upper level westerlies carry high altitude disturbances over the humid air where they trigger the giant thunderstorms which characterize springtime American weather. Afterwards the westerlies push the humidity and cloud eastward for a short time, returning dry and sunny skies to the Plains for a few days. Inexorably, southerly flow begins again and the Gulf moisture returns to fuel yet another disturbance.

Over northern sections, along the Great Lakes, New England States and over southern Canada, occasional polar air masses make a sortie southward. Marked by low pressure systems, and strong warm and cold fronts, these systems come with extensive cloud shields and usually rain. The lows pause briefly along the west slopes of the Appalachians, gathering resources to push over the peaks and into New England. Its a cloudy area, because of the lingering polar lows, the mountains, the Atlantic Ocean and the Gulf moisture. Figure 11 shows that the frequency of good eclipse weather drops below 30% from West Virginia to Maine, and again over Nova Scotia. The cloudiest area of all lies over New Hampshire and Vermont.

Since thunderstorms usually form in the afternoon and evening, a morning eclipse will escape the greatest threat of heavy weather. Cloud cover also increases during the day, reaching a maximum in the late afternoon. Figures 10 and 11 are drawn from morning cloud statistics, better reflecting the climatology that applies to the eclipse. Humid Gulf air masses also bring hazy polluted skies to areas east of the Mississippi. However, it's not likely to be much of a problem for an annular eclipse.

In western regions [Texas and Oklahoma], the dust storm season ends by May 10. Strong surface winds may whip up the dirt in small areas downwind from bare fields, but growing crops will bind the soil and tame more widespread storms. May is the month for tornadoes and hail across much of the mid west and through the Great Lakes, but is still a month or more away from the start of the hurricane season.

STRATEGIES TO COPE WITH THE WEATHER

Outside of the southwestern U. S. and Mexico, weather becomes much more variable. Cloudiness changes with each passing high and low. Fortunately North America is blessed with a profusion of weather forecasting services. Forecasts suitable for initial planning should be available 5 to 6 days before the eclipse. By the third day - May 7 - forecasters should be able to zero in more accurately and chasers can begin to plan their final site. Look for a dry westerly flow behind a strong cold front, if possible. In May 1984 those conditions brought clear skies across the southern states and a fine annular eclipse was enjoyed by millions from Georgia to the Carolina's. The motion of these systems can be forecast quite accurately about 36 to 48 hours ahead of time, allowing lots of time for planning and travel.

If weather systems expected for eclipse day are weak or poorly defined, forecasts will be less accurate and inclined to change as the day approaches. Chasers may have to travel over greater distances to reach areas where forecasters are more certain of events on eclipse day. Staying closer to home and using the satellite imagery shown on commercial television to plan at short notice is another alternative. A few hours driving may place you in an opening in an otherwise unpromising sky. Be careful when using such images, since many television stations process the pictures for a more attractive display and may lose some of the finer and smaller cloud details. The larger weather systems with the greatest amount of cloud are always visible. Watch the various forecasts from several different channels, or contact the U. S. National Weather Service or Environment Canada to allay any doubts.

12

THE AZORES

A few rocky islands in the eastern Atlantic offer a landsman's perch from which to watch the eclipse as it speeds to Africa. The islands are located under a strong and permanent high pressure anticyclone, but skies are anything but clear in the humid air. Santa Maria Island shows only a 15% frequency of scattered cloud at eclipse time. Clear skies have a frequency of zero! However, this altogether dismal statistic may not be entirely representative of the eclipse prospects. Oceanic islands often have a considerable variation in cloudiness between leeward and windward sides. In this case the improvement is probably meager. Table 15 shows that none of the sites in the Azores have promising statistics.

MOROCCO

As the golden-ringed sun settles toward the horizon it reaches the shores of Morocco and the city of Casablanca. It's not an auspicious ending. Casablanca is a fairly cloudy city, with a climate that resembles that in the Baja. As in Mexico, the controlling influence on cloudiness in northwest Africa is a large and permanent high pressure cell - the same one which controls the weather of the Azores. Strong northerly winds circulating around this high push the cool Canary Current southward along the coast of Morocco. The by now familiar combination of cool ocean breezes and moist air trapped beneath a temperature inversion conspire to cloak the coast in low clouds and occasional fogs. The eclipse ends in much the same weather as it began.

Inland regions usually escape the marine cloudiness, especially on the rising slopes of the towering Atlas mountains. The probability of a sunny sunset increases from a dismal 30% on the Atlantic shores to a promising 50% near the end of the track (Figure 9). Cloud seems to pile up against the mountains in the neighborhood of Ifrane, which reports fewer sunshine hours than any other Moroccan location in Table 15. Cloudiness in the interior comes from a variety of sources. The most likely is a low pressure depression related to more intense systems moving over Europe. Another is a passing low pressure disturbance which travels eastward into the Mediterranean. Sometimes the latter originate in the Atlantic, and pass through the Straits of Gibraltar. At other times they form on the east side of the Atlas Mountains and move across the northern reaches of the Sahara Desert through Algeria and Libya.

If the track is just right, these lows can draw hot oppressive dusty air from the desert, a wind known as the sahat in Morocco and the scirocco elsewhere in north Africa. The scirocco is not usually as intense over Morocco as other parts of the Mediterranean, but the dust can be a considerable problem for eclipse chasers. One description, though not over Morocco, confesses to a yellowish leaden sky "through which the sun can be seen only as pale disk, if at all." Satellites have traced Saharan dust all the way to the Caribbean. From Tangier eastward about 5 scirocco days per month are reported in May.

As the scirocco lows move eastward, winds turn northerly again and moist air is drawn inland against the Atlas Mountains. Lifted by the terrain, clouds build deeply, bringing the occasional rains and thundershowers of spring. These variable weather systems are especially favored in May. Even so, sunshine is the dominant element, with most inland areas receiving an amount comparable to northern Mexico.

One of the major problems for sunset eclipses is the apparent thickening of cloud near the horizon due to perspective effects. Skies must be fairly clear to hold promise of an open horizon, as many discovered in southern California in 1992. Haze may obscure the horizon, hiding low cloud layers until silhouetted by the declining sun, leaving only a half hour or less for movement to a more promising location. Since the terminus of the eclipse is on the Sahara side of the Atlas Mountains, the horizon view to the west will be blocked by the terrain. These are significant peaks with some summits reaching over 3500 meters along the center line. The most promising chances are likely to be found about 150 kilometers inland, perhaps somewhere along the road joining Marrakech and Meknes where the view to the setting sun is unobstructed. Another promising route is along the highway from Casablanca to Khouribga and beyond, climbing steadily upward from the coastal plain.

Chasers will have the best chance of success if they've scouted a few locations beforehand. Up-to-date weather information and a commitment to mobility will also help. Luckily the eclipse is late in the day, affording considerable time and daylight for planning and decisions.

OBSERVING THE ECLIPSE

EYE SAFETY DURING SOLAR ECLIPSES

The Sun can be viewed safely with the naked eye only during the few brief minutes of a *total* solar eclipse. Annular and partial solar eclipses are *never* safe to watch without taking special precautions. Although more than 88% of the Sun's surface is obscured during May's annular phase, the remaining photospheric annulus is intensely bright and cannot be viewed directly without eye protection. *Do not attempt to observe the partial or annular phases of the eclipse with the naked eye. Failure to use appropriate filtration may result in permanent eye damage or blindness!*

Generally, the same equipment, techniques and precautions used to observe the Sun outside of eclipse are required [Chou, 1981; Marsh, 1982]. There are several safe methods which may be used to watch the partial and annular phases. The safest of these is projection, in which a pinhole or small opening is used to cast the image of the sun on a screen placed a half-meter or more beyond the opening. Projected images of the sun may even be seen on the ground in the small openings created by interlacing fingers, or in the dappled sunlight beneath a tree. Binoculars can also be used to project a magnified image of the sun on a white card, but you must avoid the temptation of using these instruments for direct viewing.

Direct viewing of the sun should only be done using filters specifically designed for this purpose. Such filters usually have a thin layer of aluminum, chromium or silver deposited on their surfaces which attenuates both the visible and the infrared energy. Experienced amateur and professional astronomers may use one or two layers of completely exposed and fully developed black-and-white film, provided the film contains a silver emulsion. Since developed color films lack silver, they are unsafe for use in solar viewing. A widely available alternative for safe eclipse viewing is a number 14 welder's glass. However, only Mylar or glass filters specifically designed for the purpose should used with telescopes or binoculars.

Unsafe filters include color film, smoked glass, photographic neutral density filters and polarizing filters. Deep green or grey filters often sold with inexpensive telescopes are also dangerous. They should not be used for viewing the sun at any time since they often crack from overheating. Do not experiment with other filters unless you are certain that they are safe. Damage to the eyes comes predominantly from invisible infrared wavelengths. The fact that the sun appears dark in a filter or that you feel no discomfort does not guarantee that your eyes are safe. Avoid all unnecessary risks. Your local planetarium or amateur astronomy club is a good source for additional information.

SKY AT MAXIMUM ECLIPSE

Since annular eclipses are not accompanied by the twilight skies seen during total eclipses, they do not present an especially good opportunity to view planets in the daytime sky. Nevertheless, Venus can be observed in broad daylight provided that the sky is cloud free and of high transparency (i.e. - no dust or particulates). During the May 1994 eclipse, Venus will be located 27.7° east of the Sun. Look for the planet by first covering the eclipsed Sun with an extended hand. Other planets may be attempted but chances of successful detection are quite small. The following ephemeris [using Brentagnon and Simon, 1986] gives the positions of the naked eye planets during the eclipse. **Delta** is the distance of the planet from Earth (A.U.'s), **V** is the apparent visual magnitude of the planet, and **Elong** gives the solar elongation or angle between the Sun and planet. Note that Jupiter is near opposition and will be below the horizon during the eclipse for all observers.

```
Planetary Ephemeris: 10 May 1994 17:00:00 UT    Equinox of Mean Date

Planet       RA      Declination    Delta    V    Diameter Phase    Elong

Sun       3ʰ 9ᵐ25ˢ   17°41'14"     1.0099  -26.7  1900.5"    -        -
Mercury   3 57  1    21 56 39      1.2204   -1.2     5.5    0.87   12.0°E
Venus     5  5 20    23 58 50      1.4370   -3.4    11.6    0.88   27.7°E
Mars      1 15 15     6 58 12      2.1730    1.4     4.3    0.97   29.8°W
Jupiter  14 26 11   -13 -1 -5      4.4349   -2.0    44.4    1.00  168.6°E
Saturn   22 51 57    -8-57-48     10.0626    0.4    16.4    1.00   68.9°W
```

ECLIPSE PHOTOGRAPHY

The eclipse may be safely photographed provided that the above precautions are followed. Almost any kind of camera with manual controls can be used to capture this rare event. However, a lens with a fairly long focal length is recommended to produce as large an image of the Sun as possible. A standard 50 mm lens yields a minuscule 0.5 mm image, while a 200 mm telephoto or zoom produces a 1.9 mm image. A better choice would be one of the small, compact catadioptic or mirror lenses which have become widely available in the past ten years. The focal length of 500 mm is most common among such mirror lenses and yields a solar image of 4.6 mm. Adding 2x tele-converter will produce a 1000 mm focal length which doubles the Sun's size to 9.2 mm. Focal lengths in excess of 1000 mm usually fall within the realm of amateur telescopes. If full disk eclipse photography on 35 mm format is planned, the focal length of the telescope or lens must be 2600 mm or less. Longer focal lengths will only permit photography of a portion of the Sun's disk. For any particular focal length, the diameter of the Sun's image is approximately equal to the focal length divided by 109.

A mylar or glass solar filter must be used on the lens at all times for both photography and safe viewing. Such filters are most easily obtained through manufacturers and dealers listed in *Sky & Telescope* and *Astronomy* magazines. These filters typically attenuate the Sun's visible and infrared energy by a factor of 100,000. However, the actual filter attenuation and choice of ISO film speed will play critical roles in determining the correct photographic exposure. A low to medium speed film is recommended (ISO 50 to 100) since the Sun gives off abundant light. The easiest method for determining the correct exposure is accomplished by running a calibration test on the uneclipsed Sun. Shoot a roll of film of the mid-day Sun at a fixed aperture [f/8 to f/16] using every shutter speed between 1/1000 and 1/4 second. After the film is developed, the best exposures are noted and may be used to photograph the partial and annular phases since the Sun's surface brightness remains constant throughout the eclipse.

Another interesting way to photograph the eclipse is to record its various phases all on one frame. This is accomplished by using a stationary camera capable of making multiple exposures (check the camera instruction manual). Since the Sun moves through the sky at the rate of 15 degrees per hour, it slowly drifts through the field of view of any camera equipped with a normal focal length lens (i.e. - 35 to 50 mm). If the camera is oriented so that the Sun drifts along the frame's diagonal, it will take over three hours for the Sun to cross the field of a 50 mm lens. The proper camera orientation can be determined through trial and error several days before the eclipse. This will also insure that no trees or buildings obscure the camera's view during the eclipse. The Sun should be positioned along the eastern (left) edge or corner of the viewfinder shortly before the eclipse begins. Exposures are then made throughout the eclipse at five minute intervals. The camera must remain perfectly rigid during this period and may be to clamped to a wall or fence post since tripods are easily bumped. The final photograph will consist of a string of Suns, each showing a different phase of the eclipse.

Finally, an eclipse effect which is easily captured with point-and-shoot or automatic cameras should not be overlooked. During the eclipse, the ground under nearby shade trees is covered with small images of the crescent Sun. The gaps between the tree leaves act like pinhole cameras and each one projects its own tiny image of the Sun. The effect can be duplicated by forming a small aperture with one's hands and watching the ground below. The pinhole camera effect becomes more prominent with increasing eclipse magnitude. Virtually any camera can be used to photograph the phenomenon, but automatic cameras must have their flashes turned off since this will obliterate the pinhole images.

For more information on eclipse photography, observations and eye safety, see FURTHER READING in the BIBLIOGRAPHY.

CONTACT TIMINGS FROM THE PATH LIMITS

Precise timings of second and third contacts, made near the northern and southern limits of the umbral path (i.e. - the graze zones), are of value in determining the diameter of the Sun relative to the Moon at the time of the eclipse. Such measurements are essential to an ongoing project to monitor changes in the solar diameter. Due to the conspicuous nature of the eclipse phenomena and their strong dependence on geographical location, scientifically useful observations can be made with relatively modest equipment. Inexperienced observers are cautioned to use great care in making such observations. The safest timing technique consists of the inspection of a projected image of the rather than direct viewing of the solar disk. The observer's geodetic coordinates are required and can be measured from USGS or other large scale maps. If a map is unavailable, then a detailed description of the observing site should be included which provides information such as distance and directions of the nearest towns/settlements, nearby landmarks, identifiable buildings and road intersections. The method of contact timing should also be described, along with an estimate of the error. The precisional requirements of these observations are ±0.5 seconds in time, 1" (~30 meters) in latitude and longitude, and ±20 meters (~60 feet) in elevation. The International Occultation Timing Association (IOTA) coordinates observers world-wide during each eclipse. For more information and submission of graze observations, write to:

> International Occultation Timing Association
> Dr. David W. Dunham
> 1177 Collins Ave., SW
> Topeka, KS 66604
> U. S. A.

PLOTTING THE PATH ON MAPS

If high resolution maps of the umbral path are needed, the coordinates listed in Table 7 are conveniently provided at 1° increments of longitude to assist plotting by hand. The path coordinates in Table 3 define a line of maximum eclipse at five minute increments in Universal Time. It is also advisable to include lunar limb corrections to the northern and southern limits listed in Table 6, especially if observations are planned from the graze zones. Global Navigation Charts (1:5,000,000), Operational Navigation Charts (scale 1:1,000,000) and Tactical Pilotage Charts (1:500,000) of many parts of the world can be obtained from the Defense Mapping Agency. For specific information about map availability, purchase prices, and ordering instructions, call DMA at 1-800-826-0342 (USA) or (301) 227-2495 (outside USA). The address is:

> Defense Mapping Agency CSC
> Attn: PMA
> Washington, DC 20315-0010, USA.

Topographic maps of the United States at various scales (1:24,000, 1:62,500, 1:100,000, 1:250,000) can be ordered from:

> Branch of Distribution
> U. S. Geological Survey
> 1200 South Eads Street
> Arlington, Virginia 22202, U. S. A.

It's also advisable to check the telephone directory for any map specialty stores in your city or metropolitan area. They often have large inventories of many maps which available for immediate delivery.

ALGORITHMS, EPHEMERIDES AND PARAMETERS

Algorithms for the eclipse predictions were developed Espenak primarily from the *Explanatory Supplement* [1974] with additional algorithms from Meeus, Grosjean and Vanderleen [1966]. The solar and lunar ephemerides were generated from the JPL DE200 and LE200, respectively. All eclipse calculations were made using a value for the Moon's radius of $k=0.2722810$ for umbral contacts, and $k=0.2725076$ [adopted IAU value] for penumbral contacts. Center of mass coordinates were used except where noted. An extrapolated value for ΔT of 59.5 seconds was used to convert the predictions from Terrestrial Dynamical Time to Universal Time.

The primary source for geographic coordinates used in the local circumstances tables is *The New International Atlas* (Rand McNally, 1991). Elevations for major cites were taken from *Climates of the World* (U. S. Dept. of Commerce, 1972).

ACKNOWLEDGMENTS

Most of the predictions presented in this publication were generated on a Macintosh IIfx. Additional computations, particularly those dealing with Watts' datum and the lunar limb profile were performed on a DEC VAX 11/785 computer. Word processing and page layout for the publication were done on a Macintosh using Microsoft Word v5.1. Figure annotation was done with Claris MacDraw Pro.

We thank Francis Reddy who helped develop the data base of geographic coordinates for major cities used in the local circumstances predictions. Dr. Wayne Warren graciously provided a draft copy of the *IOTA Observer's Manual* for use in describing contact timings near the path limits. We also want to thank Dr. John Bangert for several valuable discussions and for sharing the USNO mailing list for the eclipse *Circulars*. The format and content or this work has drawn heavily upon over 40 years of eclipse *Circulars* published by the U. S. Naval Observatory. We owe a debt of gratitude to past and present staff of that institution who have performed this service for so many years. In particular, we would like to recognize the work of Julena S. Duncombe, Alan D. Fiala, Marie R. Lukac, John A. Bangert and William T. Harris. The support of Environment Canada is acknowledged in the acquisition and arrangement of the weather data. Finally, the authors thank Goddard's Laboratory for Extraterrestrial Physics for several minutes of CPU time on the LEPVX2 computer.

The names and spellings of countries, cities and other geopolitical regions are not authoritative, nor do they imply any official recognition in status. Corrections to names, geographic coordinates and elevations are actively solicited in order to update the data base for future eclipses. All calculations, diagrams and opinions presented in this publication are those of the authors and they assume full responsibility for their accuracy.

BIBLIOGRAPHY

REFERENCES

Brentagnon, P. and Simon, J. L., *Planetary Programs and Tables from -4000 to +2800*, Willmann-Bell, Richmond, Virginia, 1986.

Chou, B. R., "Safe Solar Filters", *Sky & Telescope*, August 1981, p. 119.

Climates of the World, U. S. Dept. of Commerce, Washington DC, 1972.

Dunham, J. B, Dunham, D. W. and Warren, W. H., *IOTA Observer's Manual*, (draft copy), 1992.

Espenak, F., *Fifty Year Canon of Solar Eclipses: 1986 - 2035*, NASA RP-1178, Greenbelt, MD, 1987.

Explanatory Supplement to the Astronomical Ephemeris and the American Ephemeris and Nautical Almanac, Her Majesty's Nautical Almanac Office, London, 1974.

Herald, D., "Correcting Predictions of Solar Eclipse Contact Times for the Effects of Lunar Limb Irregularities", *J. Brit. Ast. Assoc.*, 1983, **93**, 6.

Marsh, J. C. D., "Observing the Sun in Safety", *J. Brit. Ast. Assoc.*, 1982, **92**, 6.

Meeus, J., Grosjean, C., and Vanderleen, W., *Canon of Solar Eclipses*, Pergamon Press, New York, 1966.

Morrison, L. V., "Analysis of lunar occultations in the years 1943-1974...", *Astr. J.*, 1979, **75**, 744.

Morrison, L.V. and Appleby, G.M., "Analysis of lunar occultations - III. Systematic corrections to Watts' limb-profiles for the Moon", *Mon. Not. R. Astron. Soc.*, 1981, **196**, 1013.

The New International Atlas, Rand McNally, Chicago/New York/San Francisco, 1991.

van den Bergh, *Periodicity and Variation of Solar (and Lunar) Eclipses*, Tjeenk Willink, Haarlem, Netherlands, 1955.

Watts, C. B., "The Marginal Zone of the Moon", *Astron. Papers Amer. Ephem.*, 1963, **17**, 1-951.

FURTHER READING

Allen, D. and Allen, C., *Eclipse*, Allen & Unwin, Sydney, 1987.

Astrophotography Basics, Kodak Customer Service Pamphlet P150, Eastman Kodak, Rochester, 1988.

Brewer, B., *Eclipse*, Earth View, Seattle, 1991.

Covington, M., *Astrophotography for the Amateur*, Cambridge University Press, Cambridge, 1988.

Espenak, F., "Total Eclipse of the Sun", *Petersen's PhotoGraphic*, June 1991, p. 32.

Fiala, A. D., DeYoung, J. A. and Lukac, M. R., *Solar Eclipses, 1991-2000*, USNO Circular No. 170, U. S. Naval Observatory, Washington, DC, 1986.

Littmann, M. and Willcox, K., *Totality, Eclipses of the Sun*, University of Hawaii Press, Honolulu, 1991.

Lowenthal, J., *The Hidden Sun: Solar Eclipses and Astrophotography*, Avon, New York, 1984.

Mucke, H. and Meeus, J., *Canon of Solar Eclipses: -2003 to +2526*, Astronomisches Büro, Vienna, 1983.

North, G., *Advance Amateur Astronomy*, Edinburgh University Press, 1991.

Oppolzer, T. R. von, *Canon of Eclipses*, Dover Publications, New York, 1962.

Sherrod, P. C., *A Complete Manual of Amateur Astronomy*, Prentice-Hall, 1981.

Sweetsir, R. and Reynolds, M., *Observe: Eclipses*, Astronomical League, Washington DC, 1979.

Zirker, J. B., *Total Eclipses of the Sun*, Van Nostrand Reinhold, New York, 1984.

Annular Solar Eclipse of 10 May 1994

Figures

Figure 1: ORTHOGRAPHIC PROJECTION MAP OF THE ECLIPSE PATH

Annular Solar Eclipse of 10 May 1994

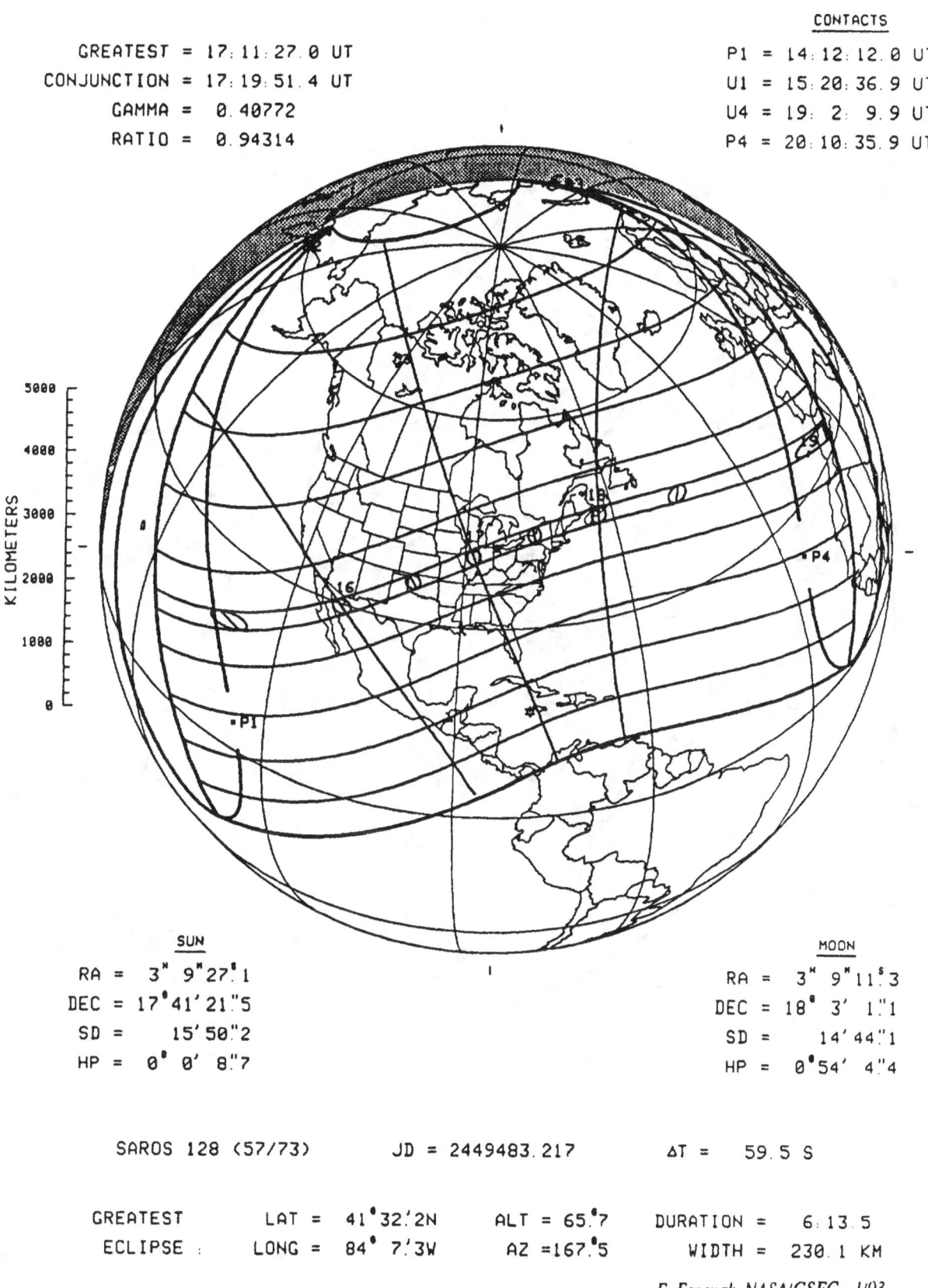

GREATEST = 17:11:27.0 UT
CONJUNCTION = 17:19:51.4 UT
GAMMA = 0.40772
RATIO = 0.94314

CONTACTS

P1 = 14:12:12.0 UT
U1 = 15:20:36.9 UT
U4 = 19:2:9.9 UT
P4 = 20:10:35.9 UT

KILOMETERS

5000
4000
3000
2000
1000
0

SUN

RA = 3ʰ 9ᵐ 27ˢ.1
DEC = 17°41'21".5
SD = 15'50".2
HP = 0° 0' 8".7

MOON

RA = 3ʰ 9ᵐ 11ˢ.3
DEC = 18° 3' 1".1
SD = 14' 44".1
HP = 0°54' 4".4

SAROS 128 (57/73)   JD = 2449483.217   ΔT = 59.5 S

GREATEST   LAT = 41°32'.2N   ALT = 65°.7   DURATION = 6:13.5
ECLIPSE   LONG = 84° 7'.3W   AZ =167°.5   WIDTH = 230.1 KM

F. Espenak, NASA/GSFC - 1/93

Figure 2: **Stereographic Projection Map of The Eclipse Path**

Annular Solar Eclipse of 10 May 1994

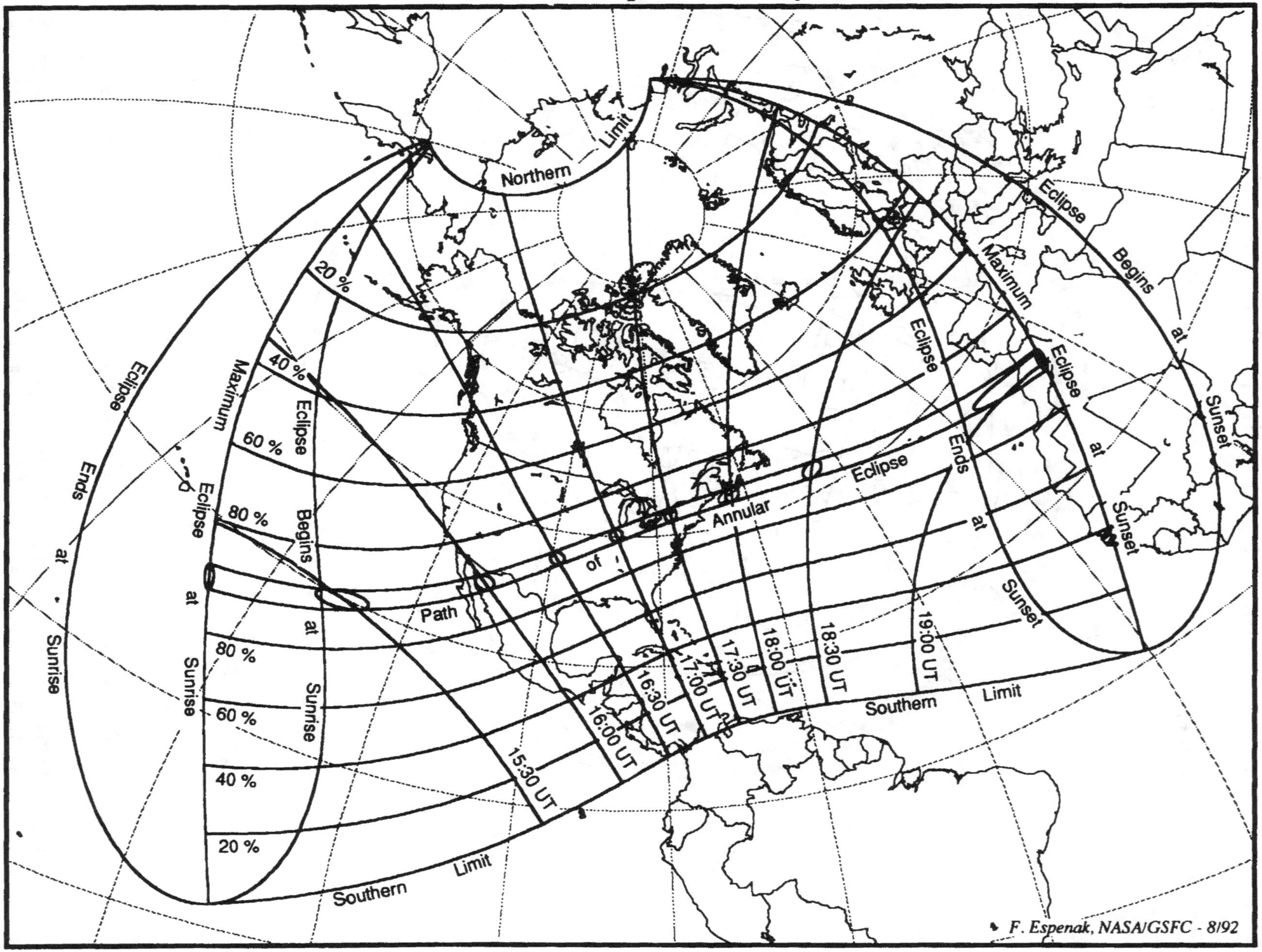

22

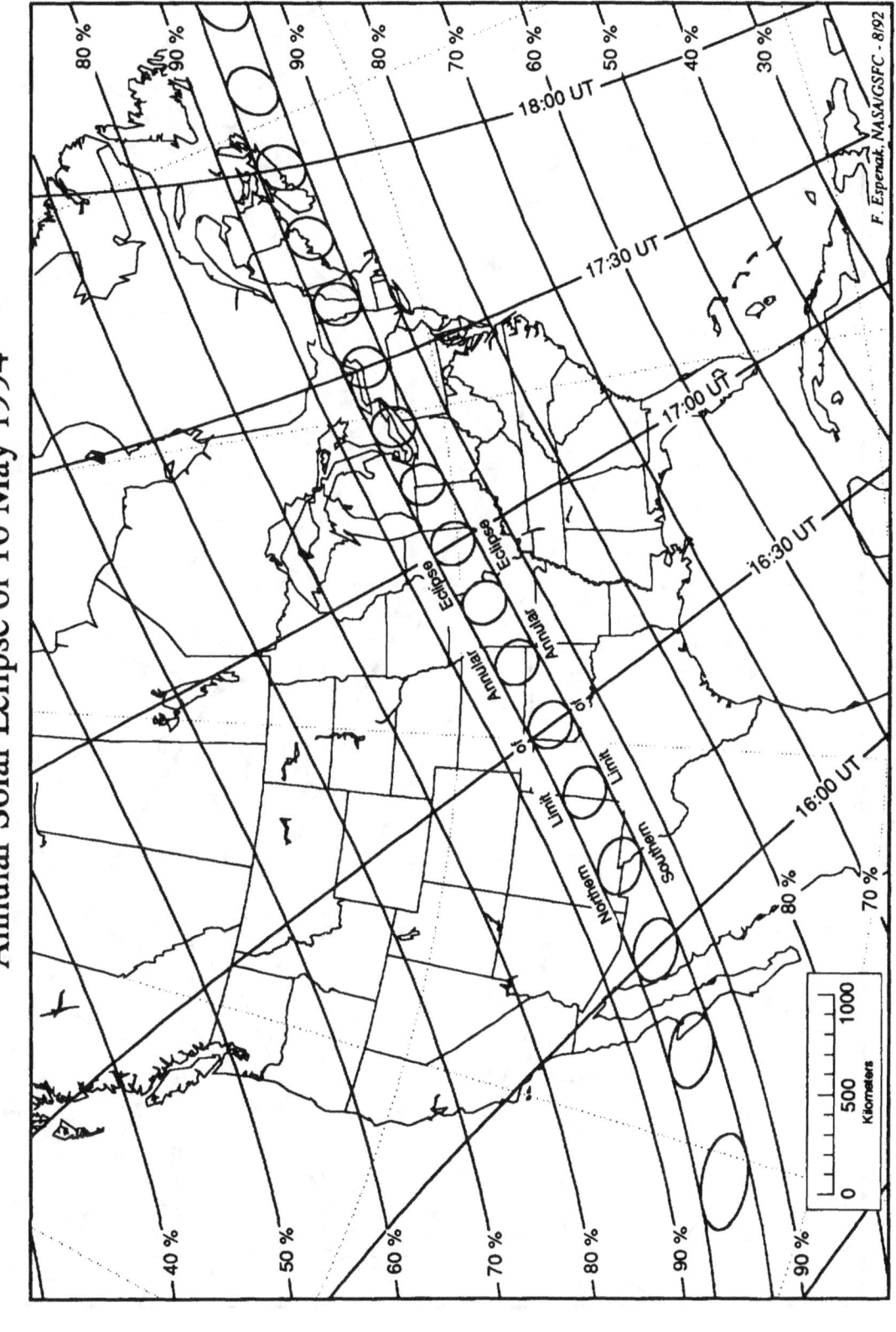

Figure 3: **THE ECLIPSE PATH IN NORTH AMERICA**

Annular Solar Eclipse of 10 May 1994

Figure 4: **The Eclipse Path in Western North America**
Annular Solar Eclipse of 10 May 1994

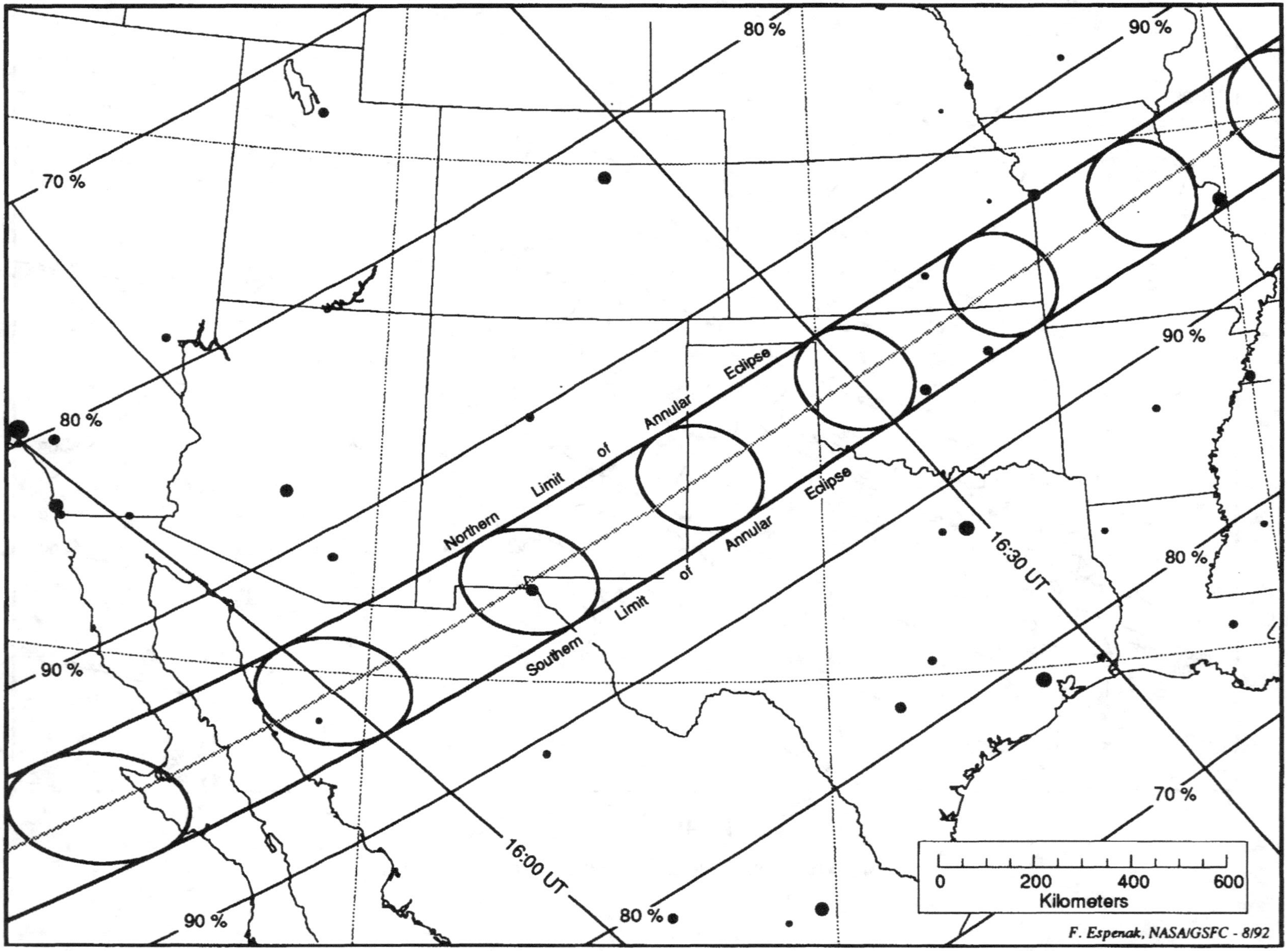

24

Figure 5: THE ECLIPSE PATH IN EASTERN NORTH AMERICA
Annular Solar Eclipse of 10 May 1994

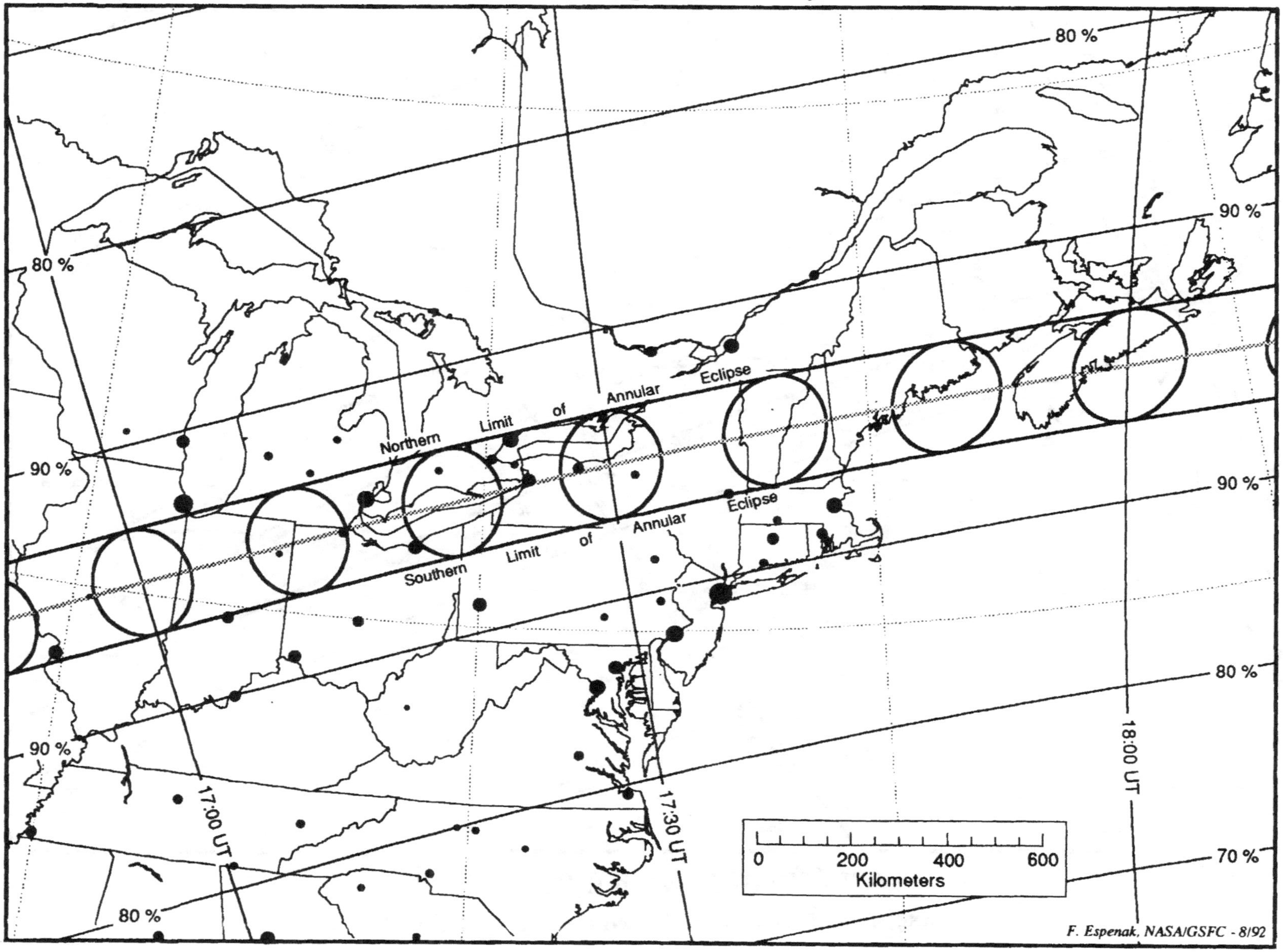

Figure 6: THE ECLIPSE PATH IN THE AZORES AND MORROCO

Annular Solar Eclipse of 10 May 1994

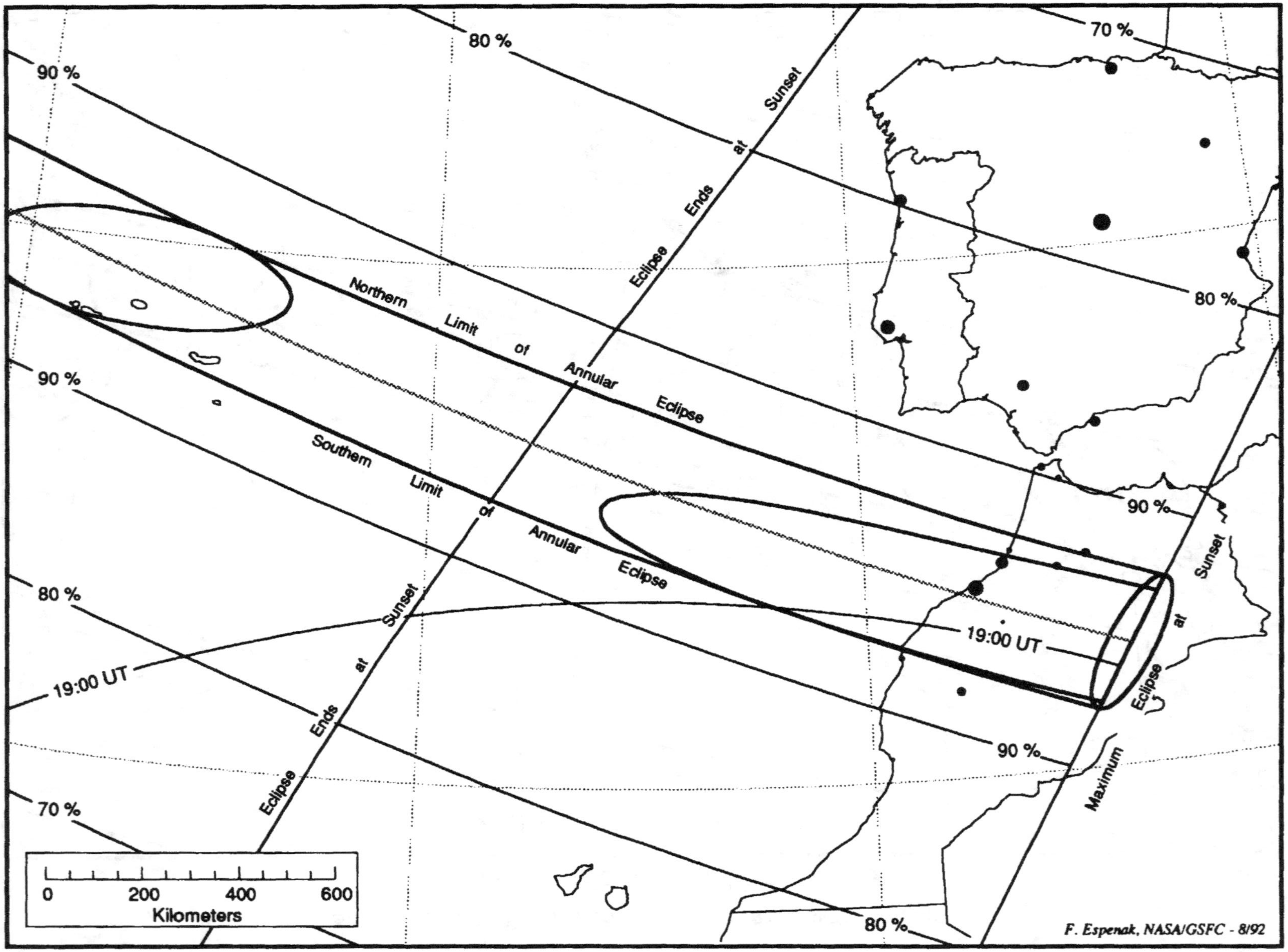

F. Espenak, NASA/GSFC - 8/92

26

Figure 7: **The Lunar Limb Profile**

Annular Solar Eclipse of 10 May 1994

L = -1.65   B = -0.12   C = -17.18

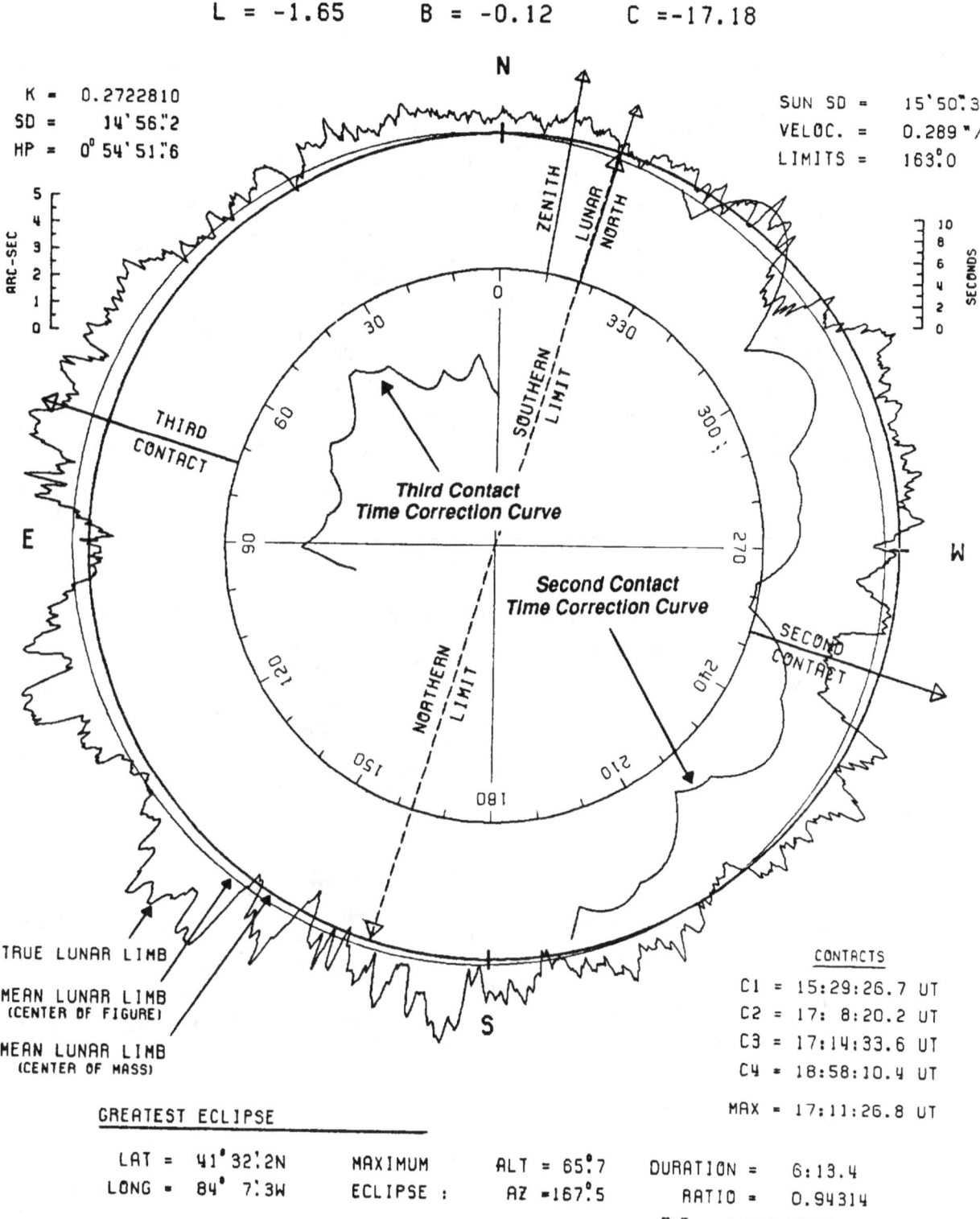

27

Figure 8: **FREQUENCY OF CLEAR SKIES DURING MAY - MEXICO**

Figure 9: **FREQUENCY OF CLEAR SKIES DURING MAY - MOROCCO**

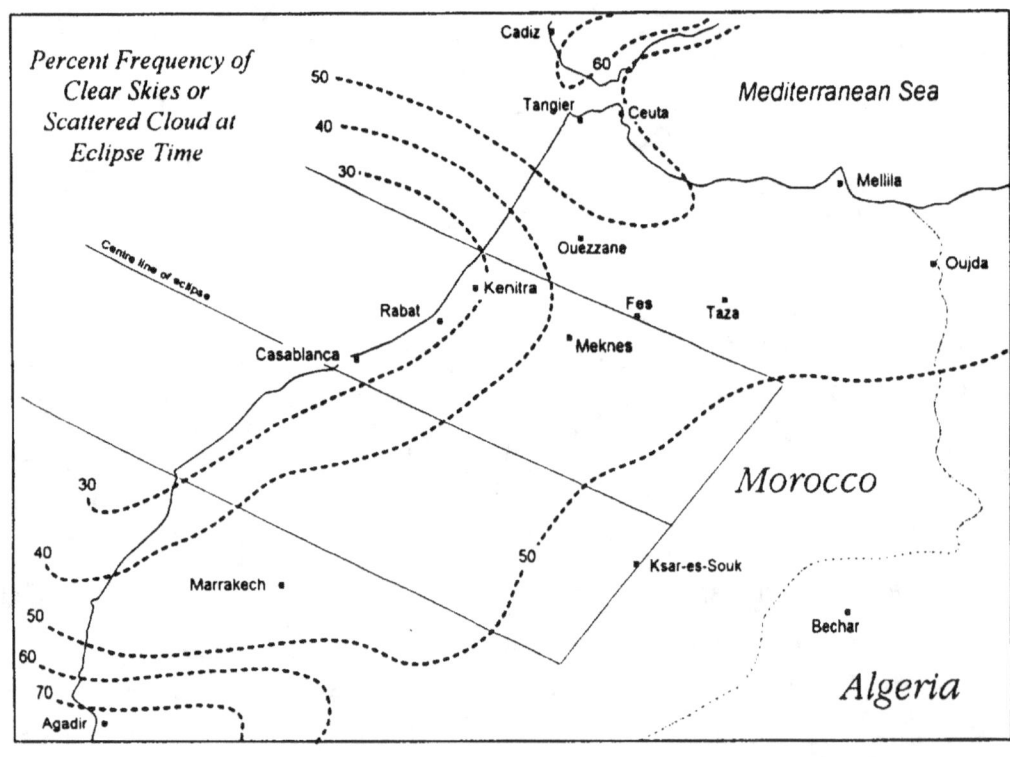

Percent Frequency
of Clear Skies or
Scattered Cloud
at Eclipse Time

Figure 11: **FREQUENCY OF CLEAR SKIES DURING MAY - EASTERN NORTH AMERICA**

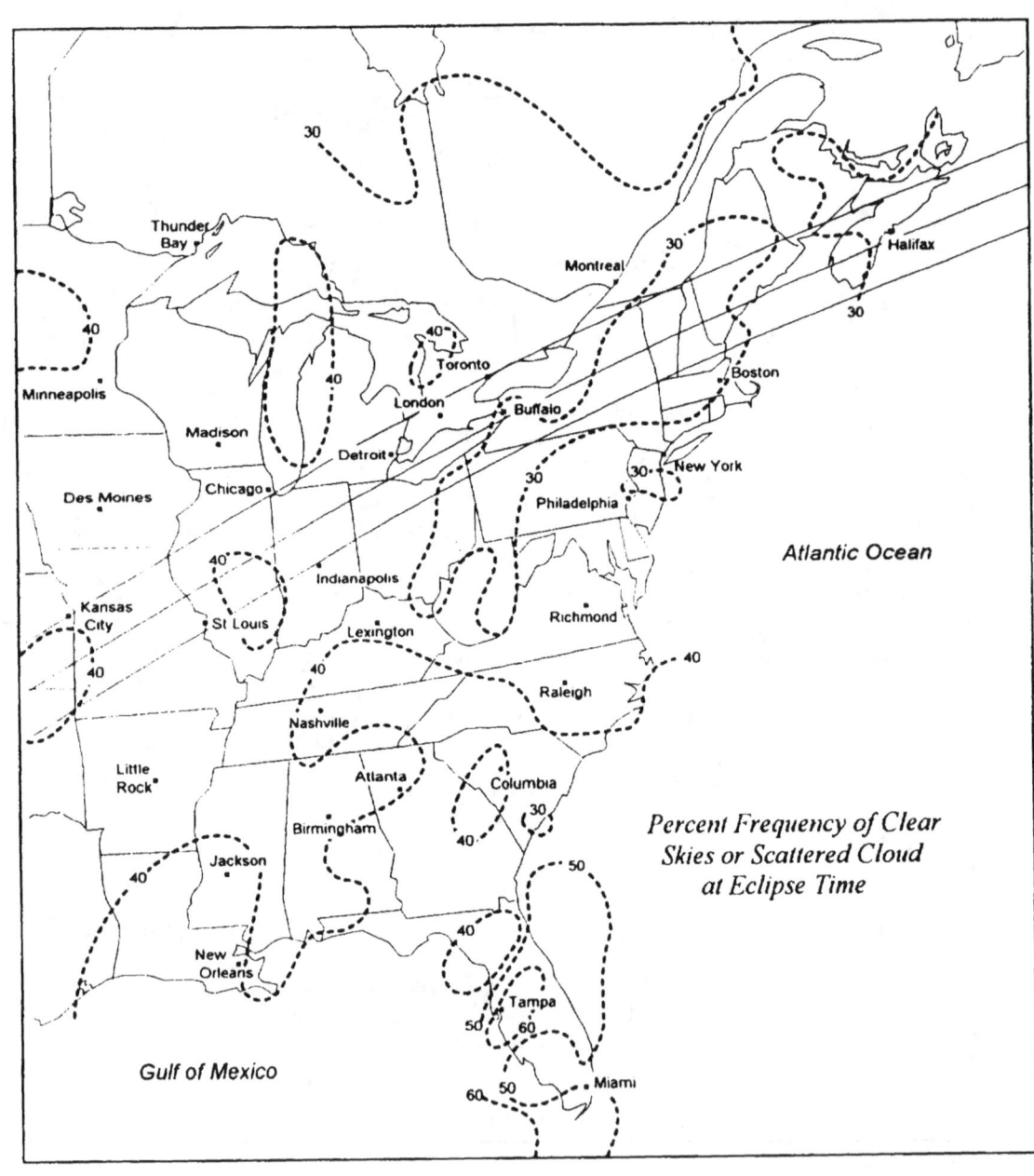

ANNULAR SOLAR ECLIPSE OF 10 MAY 1994

TABLES

Table 1

ELEMENTS OF THE ANNULAR SOLAR ECLIPSE OF 10 MAY 1994

Geocentric Conjunction   17:20:50.88 TDT   J.D. = 2449483.222811
of Sun & Moon in R.A.:   (=17:19:51.38 UT)

Instant of   17:12:26.49 TDT   J.D. = 2449483.216973
Greatest Eclipse: (=17:11:26.99 UT)

Geocentric Coordinates of Sun & Moon at Greatest Eclipse (DE200/LE200):

Sun: R.A. = 3h 9m 27.149s   Moon: R.A. = 3h 9m 11.285s
           Dec. = 17° 41′ 21.51"   Dec. = 18° 3′ 1.10"
  Semi-Diameter =   15′ 50.22"   Semi-Diameter =   14′ 44.08"
  Eq.Hor.Par. =   8.71"   Eq.Hor.Par. = 0° 54′ 4.35"
       Δ R.A. =   9.772s/h Δ R.A. =   122.984s/h
       Δ Dec. =   39.06"/h Δ Dec. =   320.89"/h

Lunar Radius   k1 = 0.2725076 (Penumbra)   Shift in Δb =   0.0"
  Constants: k2 = 0.2722810 (Umbra)   Lunar Position: Δl =   0.0"

Geocentric Libration: l = -1.6°   Brown Lun. Nbr. = 1167
(Optical + Physical)   b = -0.5°   Saros Series = 128 (57/73)
                          c = -17.2°   Ephemeris = (DE200/LE200)

Eclipse Magnitude = 0.94314   Gamma = 0.40772 ΔT =   59.5 s

Polynomial Besselian Elements for: 10 May 1994   17:00:00.0 TDT   ($=t_0$)

n	x	y	d	l_1	l_2	μ
0	-0.1734118	0.3836521	17.6861305	0.5669329	0.0206722	75.905975
1	0.4990638	0.0869394	0.0106418	-0.0000318	-0.0000317	15.001621
2	0.0000296	-0.0001183	-0.0000040	-0.0000098	-0.0000097	-0.000002
3	-0.0000056	-0.0000009				

Tan f_1 = 0.0046308   Tan f_2 = 0.0046077

At time 't_1' (decimal hours), each besselian element is evaluated by:

$$x = x_0 + x_1*t + x_2*t^2 + x_3*t^3 \quad \text{(or } x = \sum [x_n*t^n]; n = 0 \text{ to } 3)$$

where: $t = t_1 - t_0$ (decimal hours) and $t_0 = 17.000$

Note that all times are expressed in Terrestrial Dynamical Time (TDT).

Saros Series 128: Member 57 of 73 eclipses in series.

Table 2

SHADOW CONTACTS AND CIRCUMSTANCES
ANNULAR SOLAR ECLIPSE OF 10 MAY 1994

		TDT	Latitude	Ephemeris Longitude[†]	True Longitude[*]
External/Internal Contacts of Penumbra:	P1	14:13:11.3	4.951	-125.778	-125.530
	P2	16:55:59.3	56.818	165.914	166.163
	P3	17:28:40.1	70.370	70.344	70.593
	P4	20:11:35.5	23.794	-25.710	-25.461
Extreme North/South Limits of Penumbral Path:	N1	16:49: 7.9	61.992	159.981	160.230
	S1	15: 4:40.2	-17.841	-131.186	-130.938
	N2	17:35:29.7	72.160	87.594	87.843
	S2	19:20:16.2	1.069	-20.622	-20.373
External/Internal Contacts of Umbra:	U1	15:21:36.4	13.067	-145.545	-145.296
	U2	15:26:58.9	14.081	-147.232	-146.983
	U3	18:57:48.6	32.786	-3.494	-3.246
	U4	19: 3: 9.4	31.797	-5.281	-5.033
Extreme North/South Limits of Umbral Path:	N1	15:25:43.8	14.800	-147.164	-146.916
	S1	15:22:53.7	12.338	-145.622	-145.374
	N2	18:59: 3.0	33.486	-3.478	-3.230
	S2	19: 1:52.7	31.087	-5.280	-5.031
Extreme Limits of Center Line:	C1	15:24:17.3	13.563	-146.383	-146.134
	C2	19: 0:29.4	32.281	-4.396	-4.147
Instant of Greatest Eclipse:	G0	17:12:26.5	41.537	-84.369	-84.120

Circumstances at Greatest Eclipse:	Sun's Altitude = 65.7°	Path Width = 230.1 km
	Sun's Azimuth = 167.0°	Central Duration = 6m 12.7s

[†] Ephemeris Longitude is the terrestrial dynamical longitude assuming a uniformly rotating Earth.

[*] True Longitude is calculated by correcting the Ephemeris Longitude for the non-uniform rotation of Earth.
(T.L. = E.L. - 1.002738*ΔT/240, where ΔT (in seconds) = TDT - UT)

Note: Longitude is measured positive to the East.

Since ΔT is not known in advance, the value used in the predictions is an extrapolation based on pre-1992 measurements. Nevertheless, the actual value is expected to fall within ±0.3 seconds of the estimated ΔT used here.

Table 3

PATH OF THE UMBRAL SHADOW
ANNULAR SOLAR ECLIPSE OF 10 MAY 1994

Universal Time	Northern Limit Lat. ° ′	Northern Limit Long. ° ′	Southern Limit Lat. ° ′	Southern Limit Long. ° ′	Center Line Lat. ° ′	Center Line Long. ° ′	Sun Alt °	Sun Az. °	Path Width km	Central Duration m s
Limits	14 48.0	-146-54.9	12 20.3	-145-22.4	13 33.8	-146-08.1	0.0	71.8	310.6	4 34.1
15:25	15 57.0	-143-24.8	16 35.6	-133-53.8	16 37.6	-137-28.3	9.3	74.2	303.1	4 44.2
15:30	20 17.9	-132-07.1	19 26.7	-127-27.4	19 54.1	-129-39.0	18.4	77.3	294.3	4 55.7
15:35	22 41.6	-126-48.4	21 33.8	-123-06.2	22 8.4	-124-53.1	24.4	79.9	287.7	5 3.9
15:40	24 36.9	-122-51.4	23 20.7	-119-38.7	23 59.1	-121-12.2	29.2	82.4	281.9	5 10.9
15:45	26 16.7	-119-35.9	24 55.3	-116-42.1	25 36.1	-118-06.8	33.2	84.9	276.5	5 17.2
15:50	27 46.3	-116-46.0	26 21.0	-114-05.9	27 3.7	-115-24.1	36.9	87.5	271.6	5 23.0
15:55	29 8.4	-114-13.3	27 40.0	-111-44.0	28 24.1	-112-57.0	40.2	90.3	267.0	5 28.4
16: 0	30 24.5	-111-52.9	28 53.6	-109-32.5	29 38.9	-110-41.3	43.2	93.2	262.7	5 33.4
16: 5	31 35.7	-109-41.6	30 2.5	-107-28.9	30 48.9	-108-34.0	46.0	96.3	258.7	5 38.2
16:10	32 42.8	-107-37.1	31 7.4	-105-31.2	31 54.9	-106-33.0	48.6	99.6	255.0	5 42.6
16:15	33 46.2	-105-37.7	32 8.8	-103-38.1	32 57.3	-104-36.8	51.1	103.1	251.5	5 46.8
16:20	34 46.3	-103-42.1	33 7.0	-101-48.3	33 56.4	-102-44.2	53.3	106.9	248.3	5 50.8
16:25	35 43.4	-101-49.2	34 2.2	-100-01.0	34 52.5	-100-54.2	55.4	111.1	245.4	5 54.4
16:30	36 37.6	-99-58.2	34 54.6	-98-15.5	35 45.9	-99-06.0	57.4	115.5	242.7	5 57.8
16:35	37 29.2	-98-08.4	35 44.4	-96-31.1	36 36.6	-97-18.9	59.2	120.4	240.3	6 0.9
16:40	38 18.3	-96-19.2	36 31.6	-94-47.3	37 24.7	-95-32.5	60.7	125.7	238.2	6 3.7
16:45	39 4.9	-94-29.9	37 16.4	-93-03.6	38 10.4	-93-46.0	62.1	131.3	236.2	6 6.1
16:50	39 49.1	-92-40.2	37 58.8	-91-19.5	38 53.7	-91-59.2	63.3	137.4	234.6	6 8.3
16:55	40 30.9	-90-49.6	38 38.9	-89-34.7	39 34.6	-90-11.6	64.3	144.0	233.1	6 10.1
17: 0	41 10.3	-88-57.7	39 16.5	-87-48.9	40 13.1	-88-22.8	65.0	150.8	231.9	6 11.5
17: 5	41 47.2	-87-04.2	39 51.8	-86-01.6	40 49.2	-86-32.4	65.5	158.0	231.0	6 12.6
17:10	42 21.8	-85-08.6	40 24.7	-84-12.6	41 22.9	-84-40.2	65.7	165.4	230.3	6 13.3
17:15	42 53.8	-83-10.6	40 55.2	-82-21.5	41 54.1	-82-45.7	65.6	172.8	229.7	6 13.6
17:20	43 23.3	-81-09.9	41 23.1	-80-28.2	42 22.8	-80-48.8	65.3	180.2	229.5	6 13.6
17:25	43 50.2	-79-06.2	41 48.5	-78-32.2	42 49.0	-78-49.0	64.7	187.5	229.4	6 13.1
17:30	44 14.2	-76-59.2	42 11.3	-76-33.2	43 12.4	-76-46.2	63.9	194.5	229.6	6 12.3
17:35	44 35.5	-74-48.4	42 31.2	-74-31.1	43 32.9	-74-39.8	62.8	201.2	230.0	6 11.0
17:40	44 53.7	-72-33.7	42 48.3	-72-25.3	43 50.6	-72-29.7	61.5	207.6	230.7	6 9.3
17:45	45 8.7	-70-14.4	43 2.4	-70-15.7	44 5.1	-70-15.4	60.0	213.7	231.6	6 7.2
17:50	45 20.3	-67-50.4	43 13.3	-68-01.8	44 16.4	-67-56.6	58.4	219.4	232.8	6 4.7
17:55	45 28.4	-65-21.0	43 20.8	-65-43.1	44 24.2	-65-32.7	56.5	224.8	234.3	6 1.8
18: 0	45 32.6	-62-45.8	43 24.7	-63-19.3	44 28.2	-63-03.4	54.5	230.0	236.0	5 58.5
18: 5	45 32.6	-60-04.0	43 24.7	-60-49.7	44 28.3	-60-27.9	52.3	235.0	238.0	5 54.8
18:10	45 28.1	-57-14.9	43 20.4	-58-13.5	44 23.9	-57-45.5	50.0	239.7	240.4	5 50.7
18:15	45 18.5	-54-17.5	43 11.5	-55-29.9	44 14.7	-54-55.2	47.5	244.3	243.1	5 46.2
18:20	45 3.3	-51-10.2	42 57.5	-52-37.7	44 0.1	-51-55.7	44.8	248.7	246.2	5 41.3
18:25	44 41.7	-47-51.3	42 37.6	-49-35.3	43 39.4	-48-45.5	41.9	253.1	249.8	5 36.0
18:30	44 12.6	-44-18.1	42 11.0	-46-20.6	43 11.6	-45-21.9	38.7	257.4	253.9	5 30.3
18:35	43 34.5	-40-26.5	41 36.4	-42-50.4	42 35.4	-41-41.5	35.3	261.7	258.6	5 24.1
18:40	42 44.9	-36-10.0	40 52.0	-38-59.8	41 48.6	-37-38.7	31.5	266.0	264.0	5 17.4
18:45	41 39.6	-31-16.6	39 54.7	-34-40.6	40 47.6	-33-03.6	27.1	270.5	270.3	5 10.0
18:50	40 9.9	-25-20.9	38 38.8	-29-36.9	39 25.6	-27-36.4	21.9	275.4	278.1	5 1.6
18:55	37 48.1	-17-00.5	36 50.9	-23-10.2	37 23.9	-20-23.1	15.1	281.0	288.3	4 51.3
Limits	33 29.1	-3-13.8	31 5.2	-5 -1.9	32 16.8	-4 -8.8	0.0	291.1	309.0	4 32.3

Table 4

PHYSICAL EPHEMERIS OF THE UMBRAL SHADOW
ANNULAR SOLAR ECLIPSE OF 10 MAY 1994

Universal Time	Center Line Latitude ° ′	Longitude ° ′	Diameter Ratio	Eclipse Obscur.	Sun Alt °	Path Az. °	Path Width km	Major Axis km	Minor Axis km	Umbral Veloc. km/s	Central Duration m s
15:23.3	13 33.8	-146-08.1	0.9296	0.8641	0.0	-	310.6	-	264.0	-	4 34.1
15:25	16 37.6	-137-28.3	0.9319	0.8684	9.3	69.9	303.1	1582.9	254.6	4.677	4 44.2
15:30	19 54.1	-129-39.0	0.9342	0.8727	18.4	66.2	294.3	777.4	245.5	2.193	4 55.7
15:35	22 8.4	-124-53.1	0.9356	0.8754	24.4	63.3	287.7	581.5	239.8	1.595	5 3.9
15:40	23 59.1	-121-12.2	0.9367	0.8774	29.2	61.4	281.9	483.8	235.4	1.300	5 10.9
15:45	25 36.1	-118-06.8	0.9376	0.8791	33.2	60.0	276.5	423.2	231.8	1.119	5 17.2
15:50	27 3.7	-115-24.1	0.9384	0.8806	36.9	59.0	271.6	381.4	228.8	0.995	5 23.0
15:55	28 24.1	-112-57.0	0.9391	0.8818	40.2	58.3	267.0	350.6	226.1	0.904	5 28.4
16: 0	29 38.9	-110-41.3	0.9397	0.8829	43.2	57.8	262.7	326.9	223.8	0.835	5 33.4
16: 5	30 48.9	-108-34.0	0.9402	0.8839	46.0	57.5	258.7	308.2	221.7	0.781	5 38.2
16:10	31 54.9	-106-33.0	0.9406	0.8848	48.6	57.4	255.0	293.0	219.9	0.738	5 42.6
16:15	32 57.3	-104-36.8	0.9411	0.8856	51.1	57.5	251.5	280.6	218.2	0.702	5 46.8
16:20	33 56.4	-102-44.2	0.9414	0.8863	53.3	57.8	248.3	270.3	216.8	0.673	5 50.8
16:25	34 52.5	-100-54.2	0.9417	0.8869	55.4	58.3	245.4	261.7	215.5	0.649	5 54.4
16:30	35 45.9	-99-06.0	0.9420	0.8874	57.4	58.9	242.7	254.6	214.4	0.628	5 57.8
16:35	36 36.6	-97-18.9	0.9423	0.8879	59.2	59.6	240.3	248.6	213.4	0.612	6 0.9
16:40	37 24.7	-95-32.5	0.9425	0.8883	60.7	60.5	238.2	243.7	212.5	0.598	6 3.7
16:45	38 10.4	-93-46.0	0.9427	0.8886	62.1	61.5	236.2	239.6	211.8	0.587	6 6.1
16:50	38 53.7	-91-59.2	0.9428	0.8889	63.3	62.6	234.6	236.4	211.2	0.578	6 8.3
16:55	39 34.6	-90-11.6	0.9430	0.8892	64.3	63.9	233.1	233.9	210.8	0.571	6 10.1
17: 0	40 13.1	-88-22.8	0.9430	0.8893	65.0	65.2	231.9	232.1	210.4	0.567	6 11.5
17: 5	40 49.2	-86-32.4	0.9431	0.8894	65.5	66.7	231.0	231.0	210.2	0.564	6 12.6
17:10	41 22.9	-84-40.2	0.9431	0.8895	65.7	68.3	230.3	230.5	210.1	0.563	6 13.3
17:15	41 54.1	-82-45.7	0.9431	0.8895	65.6	69.9	229.7	230.6	210.0	0.564	6 13.6
17:20	42 22.8	-80-48.8	0.9431	0.8895	65.3	71.7	229.5	231.3	210.1	0.567	6 13.6
17:25	42 49.0	-78-49.0	0.9431	0.8894	64.7	73.5	229.4	232.6	210.4	0.572	6 13.1
17:30	43 12.4	-76-46.2	0.9430	0.8892	63.9	75.4	229.6	234.7	210.7	0.578	6 12.3
17:35	43 32.9	-74-39.8	0.9429	0.8890	62.8	77.4	230.0	237.4	211.1	0.587	6 11.0
17:40	43 50.6	-72-29.7	0.9427	0.8887	61.5	79.4	230.7	240.9	211.7	0.599	6 9.3
17:45	44 5.1	-70-15.4	0.9425	0.8884	60.0	81.4	231.6	245.2	212.4	0.612	6 7.2
17:50	44 16.4	-67-56.6	0.9423	0.8880	58.4	83.5	232.8	250.5	213.2	0.629	6 4.7
17:55	44 24.2	-65-32.7	0.9421	0.8875	56.5	85.7	234.3	256.9	214.2	0.649	6 1.8
18: 0	44 28.2	-63-03.4	0.9418	0.8870	54.5	87.8	236.0	264.5	215.3	0.673	5 58.5
18: 5	44 28.3	-60-27.9	0.9415	0.8864	52.3	90.0	238.0	273.7	216.6	0.702	5 54.8
18:10	44 23.9	-57-45.5	0.9411	0.8857	50.0	92.2	240.4	284.8	218.0	0.737	5 50.7
18:15	44 14.7	-54-55.2	0.9407	0.8849	47.5	94.3	243.1	298.2	219.7	0.778	5 46.2
18:20	44 0.1	-51-55.7	0.9402	0.8840	44.8	96.5	246.2	314.7	221.5	0.830	5 41.3
18:25	43 39.4	-48-45.5	0.9397	0.8830	41.9	98.6	249.8	335.2	223.7	0.893	5 36.0
18:30	43 11.6	-45-21.9	0.9391	0.8819	38.7	100.6	253.9	361.5	226.1	0.975	5 30.3
18:35	42 35.4	-41-41.5	0.9384	0.8805	35.3	102.6	258.6	396.3	228.8	1.082	5 24.1
18:40	41 48.6	-37-38.7	0.9376	0.8790	31.5	104.6	264.0	444.8	232.0	1.232	5 17.4
18:45	40 47.6	-33-03.6	0.9366	0.8772	27.1	106.4	270.3	518.0	235.8	1.458	5 10.0
18:50	39 25.6	-27-36.4	0.9354	0.8750	21.9	108.2	278.1	645.3	240.6	1.851	5 1.6
18:55	37 23.9	-20-23.1	0.9337	0.8719	15.1	109.7	288.3	953.3	247.2	2.806	4 51.3
18:59.5	32 16.8	-4-08.8	0.9300	0.8648	0.0	-	309.0	-	262.4	-	4 32.3

Table 5

LOCAL CIRCUMSTANCES ON THE CENTER LINE
ANNULAR SOLAR ECLIPSE OF 10 MAY 1994

Center Line Maximum Eclipse			First Contact				Second Contact			Third Contact			Fourth Contact			
U.T.	Dur. m s	Alt °	U.T.	P °	V °	Alt °	U.T.	P °	V °	U.T.	P °	V °	U.T.	P °	V °	Alt °
15:25	4 44	9	–	–	–	–	15:22:38	251	327	15:27:23	71	147	16:37:24	70	150	26
15:30	4 56	18	14:22:24	251	321	3	15:27:33	250	324	15:32:28	70	145	16:48:44	69	147	37
15:35	5 4	24	14:24:16	251	320	9	15:32:29	249	323	15:37:33	69	143	16:58:11	68	144	44
15:40	5 11	29	14:26:36	250	319	13	15:37:25	249	321	15:42:36	69	141	17:06:54	68	142	49
15:45	5 17	33	14:29:12	250	318	16	15:42:22	249	319	15:47:39	69	139	17:15:08	68	139	54
15:50	5 23	37	14:31:59	250	317	20	15:47:19	248	318	15:52:42	68	138	17:23:01	68	135	58
15:55	5 28	40	14:34:53	249	315	23	15:52:16	248	316	15:57:45	68	136	17:30:36	68	131	61
16:00	5 33	43	14:37:54	249	314	25	15:57:14	248	314	16:02:47	68	134	17:37:55	68	126	64
16:05	5 38	46	14:41:01	249	313	28	16:02:12	248	312	16:07:50	68	132	17:44:58	68	121	67
16:10	5 43	49	14:44:12	249	312	31	16:07:09	248	310	16:12:52	68	130	17:51:47	69	114	69
16:15	5 47	51	14:47:28	249	311	33	16:12:07	249	308	16:17:54	68	128	17:58:21	69	106	70
16:20	5 51	53	14:50:48	249	310	35	16:17:05	249	306	16:22:56	69	125	18:04:42	69	98	71
16:25	5 54	55	14:54:13	249	308	37	16:22:03	249	303	16:27:58	69	122	18:10:50	70	89	72
16:30	5 58	57	14:57:41	249	307	39	16:27:02	249	300	16:32:60	69	119	18:16:44	70	81	72
16:35	6 1	59	15:01:14	249	306	41	16:32:00	249	297	16:38:01	70	116	18:22:25	71	73	71
16:40	6 4	61	15:04:51	249	305	43	16:36:59	250	293	16:43:02	70	112	18:27:55	72	66	70
16:45	6 6	62	15:08:32	250	303	45	16:41:58	250	290	16:48:04	70	108	18:33:12	72	60	69
16:50	6 8	63	15:12:18	250	301	47	16:46:57	251	285	16:53:05	71	104	18:38:18	73	55	68
16:55	6 10	64	15:16:09	250	300	49	16:51:56	251	281	16:58:06	71	99	18:43:13	73	51	66
17:00	6 12	65	15:20:05	251	298	51	16:56:55	252	276	17:03:06	72	94	18:47:57	74	48	64
17:05	6 13	66	15:24:07	251	296	52	17:01:54	252	271	17:08:07	72	88	18:52:31	75	45	62
17:10	6 13	66	15:28:14	251	294	54	17:06:54	253	266	17:13:07	73	83	18:56:56	75	43	61
17:15	6 14	66	15:32:28	252	291	56	17:11:54	253	260	17:18:07	74	78	19:01:12	76	41	59
17:20	6 14	65	15:36:50	252	289	57	17:16:54	254	255	17:23:07	74	73	19:05:19	77	40	57
17:25	6 13	65	15:41:18	253	285	58	17:21:54	255	250	17:28:07	75	68	19:09:19	77	39	55
17:30	6 12	64	15:45:55	253	282	60	17:26:54	255	245	17:33:07	76	63	19:13:10	78	38	53
17:35	6 11	63	15:50:40	254	278	61	17:31:55	256	241	17:38:06	76	59	19:16:55	79	37	51
17:40	6 9	62	15:55:35	254	274	62	17:36:56	257	237	17:43:05	77	55	19:20:33	79	37	48
17:45	6 7	60	16:00:40	255	270	62	17:41:57	257	234	17:48:04	78	52	19:24:05	80	36	46
17:50	6 5	58	16:05:55	256	265	63	17:46:58	258	230	17:53:03	78	49	19:27:31	81	36	44
17:55	6 2	57	16:11:22	256	259	63	17:51:60	259	227	17:58:01	79	46	19:30:51	81	36	42
18:00	5 58	55	16:17:01	257	254	63	17:57:01	260	225	18:02:60	80	44	19:34:06	82	35	40
18:05	5 55	52	16:22:54	258	248	63	18:02:03	260	223	18:07:58	81	42	19:37:16	82	35	37
18:10	5 51	50	16:29:01	259	243	62	18:07:05	261	221	18:12:56	81	40	19:40:20	83	35	35
18:15	5 46	47	16:35:22	260	237	61	18:12:07	262	219	18:17:53	82	39	19:43:19	84	35	32
18:20	5 41	45	16:41:59	260	232	59	18:17:10	263	218	18:22:51	83	37	19:46:12	84	35	30
18:25	5 36	42	16:48:53	261	228	57	18:22:12	263	217	18:27:48	83	36	19:48:60	85	35	27
18:30	5 30	39	16:56:06	262	224	54	18:27:15	264	216	18:32:46	84	35	19:51:41	85	35	24
18:35	5 24	35	17:03:38	263	220	51	18:32:18	265	215	18:37:42	85	35	19:54:14	86	35	21
18:40	5 17	32	17:11:32	264	217	48	18:37:22	265	214	18:42:39	85	34	19:56:38	86	35	17
18:45	5 10	27	17:19:54	265	214	43	18:42:25	266	213	18:47:35	86	33	19:58:48	87	35	13
18:50	5 2	22	17:28:53	266	212	38	18:47:30	267	212	18:52:31	87	33	20:00:38	87	36	9
18:55	4 51	15	17:38:57	267	210	30	18:52:35	267	212	18:57:26	87	32	20:01:47	87	36	3

Table 6

TOPOCENTRIC DATA AND PATH CORRECTIONS DUE TO LUNAR LIMB PROFILE

Universal Time	Moon Topo H.P. "	Moon Topo S.D. "	M:S Rel. Ang.V "/s	Topo Lib. Long °	Sun Alt. °	Sun Az. °	Path Az. °	North Limit P.A. °	North Limit Int. '	North Limit Ext. '	South Limit Int. '	South Limit Ext. '	Central Durat. Cor. s
15:25	3252.1	885.5	0.455	-0.74	9.3	74.2	68.0	161.2	3.5	-2.0	-3.3	0.2	-7.8
15:30	3260.0	887.7	0.423	-0.78	18.5	77.3	64.4	160.0	3.4	-1.9	-2.9	0.0	-8.7
15:35	3265.0	889.0	0.403	-0.83	24.4	79.9	62.2	159.4	3.1	-2.2	-2.7	0.2	-9.5
15:40	3268.8	890.1	0.387	-0.87	29.2	82.4	60.6	159.0	3.0	-2.2	-2.7	0.4	-10.0
15:45	3272.0	891.0	0.374	-0.91	33.3	84.9	59.5	158.7	2.8	-2.3	-2.8	0.5	-10.2
15:50	3274.8	891.7	0.362	-0.95	36.9	87.6	58.6	158.5	2.7	-2.2	-2.9	0.5	-10.6
15:55	3277.1	892.3	0.353	-0.99	40.2	90.3	58.0	158.3	2.7	-2.2	-3.1	0.5	-11.0
16:00	3279.2	892.9	0.344	-1.04	43.2	93.2	57.6	158.3	2.6	-2.2	-3.0	0.5	-11.3
16:05	3281.1	893.4	0.336	-1.08	46.0	96.3	57.4	158.3	2.6	-2.2	-3.0	0.5	-11.5
16:10	3282.7	893.8	0.329	-1.12	48.7	99.6	57.5	158.4	2.6	-2.2	-2.8	0.5	-11.5
16:15	3284.2	894.2	0.323	-1.16	51.1	103.1	57.7	158.5	2.6	-2.2	-2.7	0.5	-11.7
16:20	3285.5	894.6	0.317	-1.21	53.4	106.9	58.0	158.7	2.6	-2.2	-2.6	0.4	-11.9
16:25	3286.6	894.9	0.312	-1.25	55.5	111.1	58.5	158.9	2.7	-2.2	-2.5	0.4	-12.1
16:30	3287.6	895.2	0.308	-1.29	57.4	115.5	59.2	159.2	2.7	-2.2	-2.4	0.3	-12.3
16:35	3288.5	895.4	0.304	-1.34	59.2	120.4	60.0	159.5	2.9	-2.1	-2.5	0.2	-12.6
16:40	3289.3	895.6	0.300	-1.38	60.7	125.7	61.0	159.9	2.9	-2.0	-2.6	0.0	-12.7
16:45	3289.9	895.8	0.297	-1.42	62.1	131.3	62.0	160.3	3.0	-1.8	-2.7	0.0	-12.8
16:50	3290.4	895.9	0.295	-1.46	63.3	137.5	63.2	160.7	2.8	-1.7	-2.8	0.1	-12.9
16:55	3290.9	896.0	0.293	-1.51	64.3	144.0	64.5	161.2	2.9	-2.0	-2.9	0.3	-12.9
17:00	3291.2	896.1	0.291	-1.55	65.0	150.9	66.0	161.7	2.0	-2.2	-3.1	0.5	-12.8
17:05	3291.4	896.2	0.290	-1.59	65.5	158.0	67.5	162.3	2.0	-2.2	-2.8	0.5	-12.7
17:10	3291.5	896.2	0.289	-1.64	65.7	165.4	69.1	162.9	2.2	-2.2	-2.5	0.5	-12.5
17:15	3291.5	896.2	0.289	-1.68	65.6	172.8	70.8	163.5	2.1	-2.0	-2.6	0.5	-12.3
17:20	3291.4	896.2	0.289	-1.72	65.3	180.2	72.6	164.1	2.2	-2.4	-2.8	0.6	-11.8
17:25	3291.2	896.1	0.290	-1.77	64.7	187.5	74.4	164.8	2.4	-2.6	-2.9	0.7	-11.4
17:30	3290.9	896.0	0.291	-1.81	63.9	194.5	76.4	165.4	2.5	-2.7	-2.3	0.8	-10.9
17:35	3290.5	895.9	0.293	-1.85	62.8	201.2	78.4	166.1	2.7	-2.7	-2.5	0.7	-10.4
17:40	3290.0	895.8	0.295	-1.90	61.5	207.6	80.4	166.9	2.8	-2.5	-2.5	0.7	-9.7
17:45	3289.3	895.6	0.297	-1.94	60.0	213.7	82.5	167.6	3.0	-2.2	-1.7	0.9	-10.0
17:50	3288.6	895.4	0.301	-1.98	58.3	219.4	84.6	168.3	3.2	-2.3	-1.9	1.0	-9.9
17:55	3287.7	895.2	0.304	-2.03	56.5	224.9	86.7	169.0	3.5	-2.6	-1.9	1.0	-9.7
18:00	3286.7	894.9	0.309	-2.07	54.5	230.0	88.9	169.8	3.2	-2.7	-1.7	1.3	-9.4
18:05	3285.5	894.6	0.314	-2.11	52.3	235.0	91.1	170.5	1.9	-2.8	-2.0	1.3	-9.5
18:10	3284.2	894.2	0.319	-2.15	50.0	239.7	93.2	171.2	1.8	-2.6	-2.1	1.2	-9.1
18:15	3282.8	893.9	0.326	-2.20	47.5	244.3	95.4	171.9	2.8	-2.9	-2.0	1.0	-8.7
18:20	3281.1	893.4	0.333	-2.24	44.8	248.7	97.5	172.6	2.8	-3.2	-2.0	0.9	-8.1
18:25	3279.2	892.9	0.341	-2.28	41.9	253.1	99.6	173.3	2.6	-3.5	-2.3	0.9	-7.8
18:30	3277.1	892.3	0.351	-2.32	38.7	257.4	101.6	174.0	2.5	-3.7	-2.5	0.7	-7.6
18:35	3274.6	891.6	0.361	-2.37	35.3	261.7	103.6	174.7	2.7	-3.8	-2.8	0.8	-7.5
18:40	3271.7	890.9	0.374	-2.41	31.5	266.0	105.5	175.3	3.0	-3.7	-3.1	0.9	-7.4
18:45	3268.4	890.0	0.389	-2.45	27.1	270.5	107.3	176.0	2.8	-3.5	-3.0	1.0	-7.2
18:50	3264.1	888.8	0.407	-2.49	21.9	275.4	108.9	176.6	2.5	-3.1	-3.0	1.0	-7.0
18:55	3258.3	887.2	0.432	-2.53	15.1	281.0	110.4	177.2	2.9	-2.9	-3.1	1.0	-6.5

Table 7

MAPPING COORDINATES FOR THE UMBRAL PATH

Longitude	Latitude of: Northern Limit	Latitude of: Southern Limit	Latitude of: Center Line	Universal Time at: Northern Limit	Universal Time at: Southern Limit	Universal Time at: Center Line	Circumstances on the Center Line Sun Alt	Circumstances on the Center Line Sun Az.	Circumstances on the Center Line Path Width	Circumstances on the Center Line Central Durat.
° ′	° ′	° ′	° ′	h m s	h m s	h m s	°	°	km	m s
-146 00.0	15 05.5	–	13 36.4	15:24:45.1	–	15:22:27.9				
-145 00.0	15 25.1	12 26.9	13 55.9	15:24:56.7	15:21:54.0	15:23:19.4	1	72	310	4 35.2
-144 00.0	15 45.1	12 47.5	14 15.9	15:24:54.0	15:21:56.4	15:23:23.6	2	72	309	4 36.3
-143 00.0	16 05.5	13 08.2	14 36.4	15:25:04.0	15:22:01.4	15:23:30.4	3	73	308	4 37.4
-142 00.0	16 26.4	13 29.1	14 57.2	15:25:15.7	15:22:08.9	15:23:40.0	4	73	307	4 38.6
-141 00.0	16 47.7	13 50.3	15 18.6	15:25:30.3	15:22:19.1	15:23:52.4	5	73	306	4 39.8
-140 00.0	17 09.5	14 12.2	15 40.4	15:25:47.8	15:22:32.2	15:24:07.7	6	73	306	4 41.0
-139 00.0	17 31.8	14 34.5	16 02.7	15:26:08.3	15:22:48.3	15:24:26.0	8	74	305	4 42.2
-138 00.0	17 54.5	14 57.2	16 25.4	15:26:32.1	15:23:07.5	15:24:47.4	9	74	304	4 43.5
-137 00.0	18 17.7	15 20.5	16 48.6	15:26:59.0	15:23:29.8	15:25:12.0	10	74	303	4 44.9
-136 00.0	18 41.4	15 44.2	17 12.3	15:27:29.3	15:23:55.3	15:25:39.9	11	75	302	4 46.2
-135 00.0	19 05.6	16 08.4	17 36.5	15:28:02.9	15:24:24.1	15:26:11.2	12	75	301	4 47.6
-134 00.0	19 30.2	16 33.0	18 01.2	15:28:40.1	15:24:56.4	15:26:45.9	13	75	299	4 49.0
-133 00.0	19 55.3	16 58.2	18 26.3	15:29:21.0	15:25:32.3	15:27:24.2	14	76	298	4 50.5
-132 00.0	20 20.9	17 23.9	18 51.9	15:30:05.5	15:26:11.8	15:28:06.2	16	76	297	4 52.0
-131 00.0	20 47.0	17 50.0	19 18.1	15:30:53.8	15:26:55.0	15:28:52.0	17	77	296	4 53.6
-130 00.0	21 13.6	18 16.7	19 44.7	15:31:46.1	15:27:42.1	15:29:41.7	18	77	295	4 55.1
-129 00.0	21 40.6	18 43.8	20 11.8	15:32:42.3	15:28:33.2	15:30:35.3	19	78	293	4 56.8
-128 00.0	22 08.2	19 11.5	20 39.4	15:33:42.6	15:29:28.3	15:31:33.1	20	78	292	4 58.4
-127 00.0	22 36.2	19 39.6	21 07.4	15:34:47.0	15:30:27.6	15:32:34.9	22	79	291	5 00.1
-126 00.0	23 04.6	20 08.2	21 36.0	15:35:55.7	15:31:31.1	15:33:41.1	23	79	289	5 01.9
-125 00.0	23 33.5	20 37.3	22 05.0	15:37:08.7	15:32:39.0	15:34:51.6	24	80	288	5 03.7
-124 00.0	24 02.8	21 06.9	22 34.4	15:38:26.1	15:33:51.3	15:36:06.5	26	80	286	5 05.5
-123 00.0	24 32.6	21 36.9	23 04.3	15:39:47.9	15:35:08.2	15:37:25.8	27	81	285	5 07.4
-122 00.0	25 02.8	22 07.4	23 34.6	15:41:14.3	15:36:29.6	15:38:49.8	28	82	283	5 09.3
-121 00.0	25 33.3	22 38.2	24 05.3	15:42:45.1	15:37:55.7	15:40:18.3	29	83	282	5 11.3
-120 00.0	26 04.2	23 09.5	24 36.5	15:44:20.5	15:39:26.6	15:41:51.5	31	83	280	5 13.3
-119 00.0	26 35.5	23 41.2	25 07.9	15:46:00.4	15:41:02.1	15:43:29.3	32	84	278	5 15.3
-118 00.0	27 07.0	24 13.2	25 39.7	15:47:44.9	15:42:42.5	15:45:11.8	33	85	276	5 17.4
-117 00.0	27 38.8	24 45.6	26 11.8	15:49:33.9	15:44:27.7	15:46:59.0	35	86	275	5 19.5
-116 00.0	28 10.9	25 18.2	26 44.2	15:51:27.4	15:46:17.7	15:48:50.9	36	87	273	5 21.7
-115 00.0	28 43.1	25 51.2	27 16.8	15:53:25.2	15:48:12.5	15:50:47.3	37	88	271	5 23.8
-114 00.0	29 15.5	26 24.3	27 49.6	15:55:27.4	15:50:12.0	15:52:48.3	39	89	269	5 26.0
-113 00.0	29 48.1	26 57.6	28 22.5	15:57:33.8	15:52:16.1	15:54:53.7	40	90	267	5 28.2
-112 00.0	30 20.6	27 31.1	28 55.5	15:59:44.4	15:54:24.9	15:57:03.5	41	91	265	5 30.5
-111 00.0	30 53.2	28 04.6	29 28.6	16:01:58.8	15:56:38.1	15:59:17.4	43	93	263	5 32.7
-110 00.0	31 25.8	28 38.2	30 01.7	16:04:17.0	15:58:55.6	16:01:35.4	44	94	261	5 35.0
-109 00.0	31 58.2	29 11.7	30 34.7	16:06:38.7	16:01:17.3	16:03:57.3	45	96	259	5 37.2
-108 00.0	32 30.5	29 45.2	31 07.6	16:09:03.9	16:03:43.0	16:06:22.9	47	97	258	5 39.4
-107 00.0	33 02.6	30 18.5	31 40.3	16:11:32.1	16:06:12.4	16:08:51.9	48	99	256	5 41.7
-106 00.0	33 34.4	30 51.6	32 12.7	16:14:03.2	16:08:45.4	16:11:24.1	49	101	254	5 43.8
-105 00.0	34 05.9	31 24.5	32 44.9	16:16:37.0	16:11:21.7	16:13:59.3	51	102	252	5 46.0
-104 00.0	34 37.1	31 57.0	33 16.8	16:19:13.0	16:14:01.1	16:16:37.2	52	104	250	5 48.1
-103 00.0	35 07.8	32 29.1	33 48.2	16:21:51.2	16:16:43.1	16:19:17.4	53	106	249	5 50.2
-102 00.0	35 38.0	33 00.8	34 19.2	16:24:31.1	16:19:27.7	16:21:59.8	54	109	247	5 52.3
-101 00.0	36 07.7	33 32.0	34 49.6	16:27:12.5	16:22:14.4	16:24:44.0	55	111	246	5 54.2
-100 00.0	36 36.8	34 02.7	35 19.5	16:29:55.2	16:25:03.0	16:27:29.8	56	113	244	5 56.1

Table 7
MAPPING COORDINATES FOR THE UMBRAL PATH

Longitude	Latitude of:			Universal Time at:			Circumstances on the Center Line			
	Northern Limit	Southern Limit	Center Line	Northern Limit	Southern Limit	Center Line	Sun Alt	Sun Az.	Path Width	Central Durat.
° ′	° ′	° ′	° ′	h m s	h m s	h m s	°	°	km	m s
-99 00.0	37 05.3	34 32.7	35 48.8	16:32:38.8	16:27:53.1	16:30:16.7	57	116	243	5 58.0
-98 00.0	37 33.1	35 02.1	36 17.4	16:35:23.1	16:30:44.5	16:33:04.7	58	118	241	5 59.7
-97 00.0	38 00.2	35 30.8	36 45.3	16:38:07.8	16:33:36.8	16:35:53.3	59	121	240	6 01.4
-96 00.0	38 26.7	35 58.8	37 12.5	16:40:52.6	16:36:29.9	16:38:42.4	60	124	239	6 03.0
-95 00.0	38 52.3	36 26.0	37 38.9	16:43:37.5	16:39:23.3	16:41:31.5	61	127	238	6 04.4
-94 00.0	39 17.2	36 52.4	38 04.5	16:46:22.0	16:42:16.9	16:44:20.7	62	131	236	6 05.8
-93 00.0	39 41.3	37 17.9	38 29.4	16:49:06.1	16:45:10.4	16:47:09.5	63	134	235	6 07.1
-92 00.0	40 04.6	37 42.7	38 53.4	16:51:49.4	16:48:03.5	16:49:57.8	63	137	235	6 08.3
-91 00.0	40 27.1	38 06.5	39 16.5	16:54:32.0	16:50:56.1	16:52:45.4	64	141	234	6 09.3
-90 00.0	40 48.7	38 29.5	39 38.8	16:57:13.5	16:53:47.9	16:55:32.1	64	145	233	6 10.2
-89 00.0	41 09.5	38 51.5	40 00.2	16:59:53.9	16:56:38.8	16:58:17.8	65	148	232	6 11.1
-88 00.0	41 29.4	39 12.7	40 20.8	17:02:33.1	16:59:28.7	17:01:02.2	65	152	232	6 11.8
-87 00.0	41 48.5	39 33.0	40 40.4	17:05:10.9	17:02:17.2	17:03:45.4	65	156	231	6 12.4
-86 00.0	42 06.8	39 52.3	40 59.2	17:07:47.2	17:05:04.4	17:06:27.2	66	160	231	6 12.9
-85 00.0	42 24.2	40 10.8	41 17.2	17:10:22.0	17:07:50.2	17:09:07.4	66	164	230	6 13.2
-84 00.0	42 40.8	40 28.3	41 34.2	17:12:55.1	17:10:34.3	17:11:46.0	66	168	230	6 13.5
-83 00.0	42 56.5	40 45.0	41 50.4	17:15:26.6	17:13:16.8	17:14:22.9	66	172	230	6 13.6
-82 00.0	43 11.5	41 00.7	42 05.7	17:17:56.4	17:15:57.5	17:16:58.1	66	176	230	6 13.7
-81 00.0	43 25.6	41 15.6	42 20.2	17:20:24.4	17:18:36.5	17:19:31.6	65	180	229	6 13.6
-80 00.0	43 38.9	41 29.6	42 33.9	17:22:50.5	17:21:13.5	17:22:03.1	65	183	229	6 13.4
-79 00.0	43 51.4	41 42.7	42 46.7	17:25:14.9	17:23:48.7	17:24:32.9	65	187	229	6 13.2
-78 00.0	44 03.1	41 55.0	42 58.7	17:27:37.4	17:26:21.9	17:27:00.7	64	190	229	6 12.8
-77 00.0	44 14.1	42 06.4	43 09.9	17:29:58.1	17:28:53.2	17:29:26.6	64	194	230	6 12.4
-76 00.0	44 24.3	42 17.0	43 20.3	17:32:16.9	17:31:22.4	17:31:50.6	64	197	230	6 11.8
-75 00.0	44 33.7	42 26.8	43 29.9	17:34:33.8	17:33:49.7	17:34:12.6	63	200	230	6 11.2
-74 00.0	44 42.4	42 35.8	43 38.7	17:36:48.9	17:36:14.9	17:36:32.8	62	203	230	6 10.5
-73 00.0	44 50.4	42 44.0	43 46.8	17:39:02.1	17:38:38.2	17:38:50.9	62	206	231	6 09.7
-72 00.0	44 57.6	42 51.4	43 54.1	17:41:13.4	17:40:59.4	17:41:07.1	61	209	231	6 08.9
-71 00.0	45 04.2	42 58.0	44 00.7	17:43:22.9	17:43:18.5	17:43:21.4	61	212	231	6 07.9
-70 00.0	45 10.0	43 03.9	44 06.6	17:45:30.6	17:45:35.7	17:45:33.8	60	214	232	6 06.9
-69 00.0	45 15.2	43 09.0	44 11.7	17:47:36.3	17:47:50.8	17:47:44.2	59	217	232	6 05.9
-68 00.0	45 19.7	43 13.4	44 16.2	17:49:40.3	17:50:03.9	17:49:52.7	58	219	233	6 04.8
-67 00.0	45 23.5	43 17.1	44 19.9	17:51:42.4	17:52:15.0	17:51:59.3	58	222	233	6 03.6
-66 00.0	45 26.7	43 20.1	44 23.0	17:53:42.8	17:54:24.1	17:54:04.0	57	224	234	6 02.4
-65 00.0	45 29.2	43 22.4	44 25.4	17:55:41.3	17:56:31.2	17:56:06.7	56	226	235	6 01.1
-64 00.0	45 31.1	43 24.0	44 27.2	17:57:38.1	17:58:36.3	17:58:07.6	55	228	235	5 59.8
-63 00.0	45 32.4	43 24.9	44 28.3	17:59:33.0	18:00:39.4	18:00:06.6	54	230	236	5 58.4
-62 00.0	45 33.0	43 25.2	44 28.8	18:01:26.2	18:02:40.5	18:02:03.8	54	232	237	5 57.0
-61 00.0	45 33.1	43 24.8	44 28.6	18:03:17.6	18:04:39.7	18:03:59.0	53	234	238	5 55.5
-60 00.0	45 32.6	43 23.8	44 27.8	18:05:07.3	18:06:36.8	18:05:52.4	52	236	238	5 54.1
-59 00.0	45 31.4	43 22.1	44 26.4	18:06:55.2	18:08:32.1	18:07:44.0	51	238	239	5 52.6
-58 00.0	45 29.7	43 19.9	44 24.5	18:08:41.4	18:10:25.3	18:09:33.7	50	239	240	5 51.0
-57 00.0	45 27.5	43 17.0	44 21.9	18:10:25.8	18:12:16.6	18:11:21.6	49	241	241	5 49.5
-56 00.0	45 24.6	43 13.5	44 18.7	18:12:08.5	18:14:05.9	18:13:07.6	48	243	242	5 47.9
-55 00.0	45 21.2	43 09.4	44 15.0	18:13:49.5	18:15:53.3	18:14:51.7	48	244	243	5 46.3
-54 00.0	45 17.3	43 04.8	44 10.7	18:15:28.7	18:17:38.6	18:16:34.0	47	246	244	5 44.7
-53 00.0	45 12.8	42 59.6	44 05.9	18:17:06.2	18:19:22.1	18:18:14.5	46	247	245	5 43.1
-52 00.0	45 07.8	42 53.8	44 00.5	18:18:41.9	18:21:03.5	18:19:53.1	45	249	246	5 41.4
-51 00.0	45 02.3	42 47.4	43 54.6	18:20:15.9	18:22:43.0	18:21:29.8	44	250	247	5 39.8
-50 00.0	44 56.3	42 40.6	43 48.1	18:21:48.1	18:24:20.5	18:23:04.7	43	251	248	5 38.1

Table 7

MAPPING COORDINATES FOR THE UMBRAL PATH

Longitude	Latitude of:			Universal Time at:			Circumstances on the Center Line			
	Northern Limit	Southern Limit	Center Line	Northern Limit	Southern Limit	Center Line	Sun Alt	Sun Az.	Path Width	Central Durat.
° ′	° ′	° ′	° ′	h m s	h m s	h m s	°	°	km	m s
-49 00.0	44 49.8	42 33.2	43 41.2	18:23:18.6	18:25:56.0	18:24:37.8	42	253	250	5 36.4
-48 00.0	44 42.7	42 25.2	43 33.7	18:24:47.3	18:27:29.5	18:26:08.9	41	254	251	5 34.7
-47 00.0	44 35.2	42 16.8	43 25.7	18:26:14.3	18:29:01.1	18:27:38.2	40	255	252	5 33.1
-46 00.0	44 27.2	42 07.9	43 17.3	18:27:39.5	18:30:30.6	18:29:05.5	39	257	253	5 31.4
-45 00.0	44 18.8	41 58.4	43 08.3	18:29:02.9	18:31:58.0	18:30:31.0	38	258	254	5 29.7
-44 00.0	44 09.8	41 48.5	42 58.9	18:30:24.5	18:33:23.5	18:31:54.5	37	259	256	5 28.0
-43 00.0	44 00.5	41 38.1	42 49.0	18:31:44.3	18:34:46.9	18:33:16.2	37	260	257	5 26.3
-42 00.0	43 50.6	41 27.3	42 38.7	18:33:02.2	18:36:08.2	18:34:35.8	36	261	258	5 24.6
-41 00.0	43 40.4	41 16.0	42 27.9	18:34:18.4	18:37:27.5	18:35:53.6	35	262	259	5 23.0
-40 00.0	43 29.7	41 04.2	42 16.7	18:35:32.6	18:38:44.7	18:37:09.4	34	264	261	5 21.3
-39 00.0	43 18.6	40 52.0	42 05.0	18:36:45.0	18:39:59.8	18:38:23.1	33	265	262	5 19.6
-38 00.0	43 07.0	40 39.4	41 53.0	18:37:55.6	18:41:12.7	18:39:34.9	32	266	263	5 18.0
-37 00.0	42 55.1	40 26.4	41 40.5	18:39:04.2	18:42:23.6	18:40:44.7	31	267	265	5 16.4
-36 00.0	42 42.8	40 13.0	41 27.6	18:40:11.0	18:43:32.3	18:41:52.5	30	268	266	5 14.7
-35 00.0	42 30.1	39 59.2	41 14.4	18:41:15.8	18:44:38.9	18:42:58.3	29	269	268	5 13.1
-34 00.0	42 17.0	39 45.1	41 00.8	18:42:18.7	18:45:43.3	18:44:02.0	28	270	269	5 11.5
-33 00.0	42 03.6	39 30.5	40 46.8	18:43:19.6	18:46:45.5	18:45:03.6	27	271	270	5 09.9
-32 00.0	41 49.8	39 15.7	40 32.5	18:44:18.6	18:47:45.6	18:46:03.2	26	272	272	5 08.3
-31 00.0	41 35.7	39 00.4	40 17.8	18:45:15.6	18:48:43.5	18:47:00.7	25	272	273	5 06.8
-30 00.0	41 21.2	38 44.9	40 02.8	18:46:10.6	18:49:39.1	18:47:56.0	24	273	275	5 05.2
-29 00.0	41 06.4	38 29.0	39 47.5	18:47:03.6	18:50:32.6	18:48:49.3	23	274	276	5 03.7
-28 00.0	40 51.3	38 12.9	39 31.8	18:47:54.6	18:51:23.8	18:49:40.5	22	275	277	5 02.2
-27 00.0	40 35.9	37 56.4	39 15.9	18:48:43.6	18:52:12.9	18:50:29.5	21	276	279	5 00.7
-26 00.0	40 20.2	37 39.7	38 59.7	18:49:30.5	18:52:59.7	18:51:16.4	20	277	280	4 59.2
-25 00.0	40 04.3	37 22.7	38 43.2	18:50:15.4	18:53:44.2	18:52:01.2	19	277	282	4 57.8
-24 00.0	39 48.1	37 05.4	38 26.4	18:50:58.3	18:54:26.6	18:52:43.8	19	278	283	4 56.4
-23 00.0	39 31.6	36 47.9	38 09.4	18:51:39.0	18:55:06.7	18:53:24.3	18	279	285	4 54.9
-22 00.0	39 14.8	36 30.1	37 52.2	18:52:17.8	18:55:44.5	18:54:02.7	17	280	286	4 53.6
-21 00.0	38 57.9	36 12.2	37 34.7	18:52:54.4	18:56:20.1	18:54:38.8	16	281	287	4 52.2
-20 00.0	38 40.7	35 54.0	37 17.1	18:53:29.0	18:56:53.5	18:55:12.9	15	281	289	4 50.8
-19 00.0	38 23.3	35 35.7	36 59.2	18:54:01.5	18:57:24.7	18:55:44.7	14	282	290	4 49.5
-18 00.0	38 05.7	35 17.1	36 41.1	18:54:31.9	18:57:53.6	18:56:14.5	13	283	292	4 48.2
-17 00.0	37 47.9	34 58.4	36 22.8	18:55:00.2	18:58:20.4	18:56:42.1	12	283	293	4 46.9
-16 00.0	37 29.9	34 39.5	36 04.4	18:55:26.6	18:58:44.9	18:57:07.5	11	284	294	4 45.6
-15 00.0	37 11.7	34 20.5	35 45.8	18:55:50.8	18:59:07.2	18:57:30.8	10	285	296	4 44.4
-14 00.0	36 53.4	34 01.4	35 27.1	18:56:13.0	18:59:27.4	18:57:52.0	9	285	297	4 43.2
-13 00.0	36 35.0	33 42.1	35 08.2	18:56:33.1	18:59:45.4	18:58:11.1	8	286	298	4 41.9
-12 00.0	36 16.4	33 22.7	34 49.2	18:56:51.1	19:00:01.2	18:58:28.0	7	287	300	4 40.8
-11 00.0	35 57.7	33 03.2	34 30.1	18:57:07.2	19:00:14.9	18:58:42.9	6	287	301	4 39.6
-10 00.0	35 38.9	32 43.6	34 10.9	18:57:21.1	19:00:26.5	18:58:55.7	5	288	302	4 38.5
-9 00.0	35 20.0	32 23.9	33 51.6	18:57:33.1	19:00:36.0	18:59:06.5	4	288	303	4 37.3
-8 00.0	35 00.9	32 04.2	33 32.2	18:57:43.1	19:00:43.4	18:59:15.2	4	289	305	4 36.2
-7 00.0	34 41.8	31 44.4	33 12.7	18:57:51.5	19:00:48.6	18:59:21.8	3	290	306	4 35.2
-6 00.0	34 22.6	31 24.4	32 53.2	18:57:54.9	19:00:52.0	18:59:26.5	2	290	307	4 34.1
-5 00.0	34 03.4	-	32 33.6	18:58:00.9	-	18:59:29.1	1	291	308	4 33.1
-4 00.0	33 44.0	-	-	18:58:49.7	-	-				
-3 00.0	-	-	-	-	-	-				

41

Table 8a
CIRCUMSTANCES AT MAXIMUM ECLIPSE ON 10 MAY 1994
FOR MEXICO

Location Name	Latitude ° '	Longitude ° '	Elev. m	U.T. h m s	Umbral Durat. m s	Path Width km	Sun Alt °	Sun Az. °	P °	V °	Eclipse Mag.	Eclipse Obs.
MEXICO												
Acapulco	16 51.0	-99-55.0	3	15:51:02.5			51	82	335	60	0.535	0.421
Aguascalientes	21 53.0	-102-18.0	—	15:56:42.3			50	88	336	53	0.688	0.597
Buenaventura	29 51.0	-107-29.0	—	16:04:34.2			47	96	337	42	0.936	0.883
Campeche	19 51.0	-90-32.0	—	16:16:44.2			66	91	335	56	0.478	0.361
Celaya	20 31.0	-100-49.0	—	15:56:29.9			51	87	335	55	0.636	0.536
Chihuahua	28 38.0	-106-05.0	1453	16:04:08.3			48	95	337	43	0.892	0.839
Ciudad Juarez	31 44.0	-106-29.0	—	16:09:44.1	5 38.4	256	49	99	337	39	0.941	0.885
Ciudad Madero	22 16.0	-97-50.0	—	16:05:21.4			56	91	335	52	0.638	0.538
Ciudad Obregon	27 29.0	-109-56.0	—	15:56:44.4			43	91	337	46	0.914	0.862
Ciudad Victoria	23 43.8	-99-09.0	—	16:05:48.2			55	93	336	49	0.690	0.600
Coatzacoalcos	18 09.0	-94-25.0	—	16:04:05.5			59	86	335	59	0.489	0.372
Cuernavaca	18 57.0	-99-13.0	—	15:56:14.3			53	85	335	57	0.576	0.468
Culiacan	24 48.0	-107-24.0	—	15:54:50.5			45	89	336	49	0.822	0.756
Durango	24 01.0	-104-40.0	—	15:57:11.0			48	90	336	50	0.769	0.693
Ensenada	31 52.0	-116-37.0	—	15:57:50.5			38	92	158	221	0.855	0.794
Guadalajara	20 40.0	-103-20.0	1704	15:52:43.9			48	86	336	54	0.674	0.580
Guaymas	27 56.0	-110-54.0	—	15:56:26.2			42	91	337	45	0.935	0.881
Hermosillo	29 04.0	-110-58.0	—	15:58:32.2	5 16.9	265	43	92	337	44	0.940	0.883
Irapuato	20 41.0	-101-21.0	—	15:55:55.8			51	87	335	54	0.647	0.549
Jalapa	19 32.0	-96-55.0	—	16:01:41.1			56	87	335	56	0.558	0.448
La Paz	24 10.0	-110-17.0	—	15:50:03.2			41	87	337	50	0.844	0.782
Leon	21 07.0	-101-40.0	—	15:56:14.8			51	88	335	54	0.662	0.566
Los Mochis	25 45.0	-108-57.0	—	15:54:38.8			44	90	337	48	0.863	0.805
Magdalena	30 38.0	-110-57.0	—	16:01:36.0	3 19.0	262	43	94	157	222	0.940	0.883
Matamoros	25 53.0	-97-30.0	—	16:13:20.6			58	98	336	45	0.720	0.635
Mazatlan	23 11.0	-106-25.0	—	15:53:05.4			46	88	336	51	0.773	0.697
Merida	20 58.0	-89-37.0	24	16:21:27.2			68	94	335	53	0.495	0.378
Mexicali	32 40.0	-115-29.0	—	16:00:29.6			39	93	158	220	0.851	0.790
Mexico City	19 24.0	-99-09.0	2408	15:57:11.4			53	86	335	56	0.586	0.479
Minatitlan	17 59.0	-94-31.0	—	16:03:32.5			59	85	335	59	0.486	0.369
Monclova	26 54.0	-101-25.0	—	16:08:07.1			53	96	336	45	0.795	0.724
Monterrey	25 40.0	-100-19.0	568	16:07:33.9			54	95	336	47	0.752	0.672
Morelia	19 42.0	-101-07.0	—	15:54:25.6			51	86	335	56	0.620	0.518
Nacozari	30 24.0	-109-39.0	—	16:02:45.8	5 32.9	261	45	95	157	222	0.940	0.883
Nuevo Loredo	27 30.0	-99-31.0	—	16:12:44.4			56	99	336	43	0.785	0.712
Orizaba	18 51.0	-97-06.0	—	15:59:58.6			56	86	335	57	0.544	0.432
Oaxaca	17 05.0	-96-43.0	—	15:57:16.4			55	83	335	60	0.495	0.379
Pachuca	20 07.0	-98-44.0	—	15:59:23.0			54	87	335	55	0.598	0.492
Poza Rica Hidalgo	20 33.0	-97-27.0	—	16:02:39.8			56	89	335	55	0.591	0.484
Puebla	19 03.0	-96-12.0	—	16:02:09.3			57	87	335	57	0.536	0.424
Queretaro	20 36.0	-100-23.0	—	15:57:23.9			52	87	335	54	0.632	0.531
Reynosa	26 07.0	-98-18.0	—	16:12:14.7			57	98	336	45	0.736	0.654
Salina Cruz	16 11.0	-95-12.0	60	15:58:33.3			57	82	334	62	0.451	0.332
Saltillo	25 25.0	-101 00.0	—	16:05:51.7			53	94	336	47	0.755	0.676
San Ignacio	27 27.0	-112-51.0	—	15:53:18.6	3 50.7	269	40	89	337	46	0.939	0.882
San Luis Potosi	22 09.0	-100-59.0	—	15:59:24.1			52	89	335	52	0.677	0.584
Santa Ana	33 38.0	-117-57.0	—	15:59:58.0			37	92	159	220	0.805	0.735
Santa Rosalia	27 19.0	-112-17.0	—	15:53:40.7			41	90	337	46	0.938	0.882
Tampico	22 13.0	-97-51.0	26	16:05:13.4			56	91	335	52	0.637	0.537
Tepic	21 30.0	-104-54.0	—	15:52:01.6			47	87	336	53	0.714	0.627
Tijuana	32 32.0	-117-01.0	—	15:58:44.0			37	92	158	221	0.837	0.773
Toluca	19 17.0	-99-40.0	—	15:56:05.3			52	86	335	56	0.590	0.484
Torreon	25 33.0	-103-26.0	—	16:02:04.9			50	93	336	47	0.789	0.717
Tuxtla Gutierrez	16 45.0	-93-07.0	—	16:04:10.1			60	84	334	62	0.436	0.316
Uruapan	19 25.0	-102-04.0	—	15:52:21.3			49	85	335	56	0.626	0.525
Veracruz	19 13.2	-96-07.2	17	16:02:39.3			57	87	335	57	0.540	0.427

Table 8b

LOCAL CIRCUMSTANCES DURING THE ANNULAR SOLAR ECLIPSE OF 10 MAY 1994 FOR MEXICO

Location Name	First Contact U.T. h m s	Alt °	P °	V °	Second Contact U.T. h m s	Alt °	P °	V °	Third Contact U.T. h m s	Alt °	P °	V °	Fourth Contact U.T. h m s	Alt °	P °	V °
MEXICO																
Acapulco	14:31:12.5	32	273	354									17:30:22.6	74	40	131
Aguascalientes	14:32:36.5	31	264	339									17:40:19.2	74	50	123
Buenaventura	14:39:55.0	29	251	316									17:45:47.7	68	66	119
Campeche	14:49:27.9	45	277	356									18:04:26.2	87	38	4
Celaya	14:32:50.7	32	267	344									17:40:00.6	76	47	123
Chihuahua	14:38:44.5	30	253	320									17:46:51.3	70	63	118
Ciudad Juarez	14:43:54.3	30	249	312	16:06:55.6	48	258	320	16:12:34.0	49	59	121	17:51:37.4	69	68	114
Ciudad Madero	14:38:21.9	36	267	342									17:52:39.2	81	47	108
Ciudad Obregon	14:34:37.5	25	252	320									17:35:27.6	65	64	127
Ciudad Victor...	14:38:15.9	35	264	337									17:53:16.0	79	51	107
Coatzacoalcos	14:40:44.6	39	276	357									17:47:56.2	84	37	123
Cuernavaca	14:33:32.2	33	271	350									17:38:59.4	77	43	125
Culiacan	14:32:01.8	26	257	328									17:35:31.6	68	58	126
Durango	14:32:59.3	29	260	332									17:40:07.3	71	55	123
Ensenada	14:40:07.5	21	243	305									17:28:59.5	57	75	134
Guadalajara	14:30:15.0	29	265	341									17:34:29.6	72	49	127
Guaymas	14:34:54.2	24	251	318									17:34:10.7	64	66	128
Hermosillo	14:36:46.3	25	250	315	15:55:54.8	42	266	333	16:01:11.7	43	50	117	17:36:11.8	64	67	124
Irapuato	14:32:23.4	31	267	343									17:39:13.1	75	47	124
Jalapa	14:37:18.4	37	272	350									17:46:31.2	81	42	119
La Paz	14:29:29.4	23	256	327									17:27:41.2	64	60	131
Leon	14:32:28.2	31	266	342									17:39:42.4	75	48	123
Los Mochis	14:32:27.9	25	255	324									17:34:05.2	66	61	128
Magdalena	14:39:31.8	26	248	312	15:59:54.0	43	195	260	16:03:13.0	44	122	186	17:39:05.2	64	70	126
Matamoros	14:43:02.2	38	262	333									18:03:11.0	80	53	86
Mazatlan	14:30:32.9	27	260	332									17:33:58.9	69	55	127
Merida	14:52:03.7	47	276	354									18:10:58.7	85	40	350
Mexicali	14:41:52.6	23	243	304									17:32:36.5	58	76	132
Mexico City	14:33:55.4	34	270	348									17:40:35.1	77	43	124
Minatitlan	14:40:28.9	39	276	357									17:47:04.0	83	37	125
Monclova	14:39:57.9	34	258	328									17:55:03.7	76	58	106
Monterrey	14:39:21.8	35	261	331									17:55:05.6	77	55	105
Morelia	14:31:44.8	31	268	346									17:36:51.0	75	45	126
Nacozari	14:39:42.3	27	249	313	15:59:59.1	44	241	306	16:05:32.0	45	76	140	17:41:39.7	65	68	124
Nuevo Loredo	14:42:51.5	36	259	328									18:01:26.4	78	57	94
Orizaba	14:36:32.9	36	273	352									17:43:46.2	80	41	123
Oaxaca	14:36:10.9	36	276	357									17:38:26.2	79	37	129
Pachuca	14:35:07.5	34	270	347									17:43:52.7	79	44	121
Poza Rica Hid..	14:37:18.3	36	270	347									17:48:27.5	81	44	116
Puebla	14:38:07.1	37	273	352									17:46:41.0	82	40	121
Queretaro	14:33:26.4	33	268	344									17:41:18.4	76	46	122
Reynosa	14:42:19.9	37	262	332									18:01:33.9	79	54	91
Salina Cruz	14:38:33.5	38	279	2									17:38:30.8	81	34	133
Saltillo	14:38:18.0	34	261	332									17:52:42.3	77	55	108
San Ignacio	14:33:20.7	22	250	318	15:51:25.0	40	294	2	15:55:15.7	40	23	92	17:28:59.9	61	66	132
San Luis Poto..	14:34:19.0	32	265	340									17:44:17.0	76	49	119
Santa Ana	14:43:20.4	21	241	301									17:29:05.7	55	79	135
Santa Rosalia	14:33:20.2	23	251	318									17:29:56.9	62	66	131
Tampico	14:38:17.4	36	267	342									17:52:27.9	81	47	108
Tepic	14:29:44.3	28	263	338									17:33:10.5	70	51	127
Tijuana	14:41:19.6	21	242	304									17:29:15.2	56	77	134
Toluca	14:33:11.7	33	270	348									17:38:59.3	77	44	125
Torreon	14:36:07.6	31	259	329									17:46:48.4	73	57	116
Tuxtla Gutier...	14:42:51.2	41	280	3									17:45:42.5	84	34	134
Uruapan	14:30:25.0	30	268	345									17:33:47.5	73	46	128
Veracruz	14:38:21.6	37	273	352									17:47:28.0	82	41	119

Table 9a
CIRCUMSTANCES AT MAXIMUM ECLIPSE ON 10 MAY 1994
FOR CENTRAL AMERICA AND CARIBBEAN

Location Name	Latitude ° ′	Longitude ° ′	Elev. m	U.T. h m s	Umbral Durat. m s	Path Width km	Sun Alt °	Sun Az. °	P °	V °	Eclipse Mag.	Eclipse Obs.
BELIZE												
Belize City	17 30.0	-88-12.0	6	16:18:07.8			68	86	335	62	0.386	0.266
Belmopan	17 15.0	-88-46.0	–	16:16:02.4			67	85	335	62	0.387	0.267
COSTA RICA												
Cartago	9 51.0	-83-55.0	–	16:15:29.7			70	65	334	85	0.126	0.052
Limon	9 59.0	-83-01.0	–	16:18:49.1			71	64	335	87	0.117	0.046
Puntarenas	9 58.0	-84-50.0	–	16:12:45.5			68	67	334	83	0.142	0.062
San Jose	9 56.0	-84-05.0	1234	16:15:05.2			70	66	334	84	0.130	0.055
EL SALVADOR												
San Miguel	13 28.0	-88-10.0	–	16:09:56.5			66	77	334	71	0.281	0.168
San Salvador	13 42.0	-89-12.0	734	16:07:36.8			64	78	334	70	0.302	0.187
Santa Ana	14 00.0	-89-33.0	–	16:07:18.8			64	78	334	69	0.315	0.198
GUATEMALA												
Antigua	14 33.0	-90-42.0	–	16:05:29.8			63	80	334	67	0.345	0.227
Guatemala City	14 38.0	-90-31.0	1593	16:06:05.4			63	80	334	67	0.345	0.226
Mazatenango	14 31.0	-91-30.0	–	16:03:29.2			61	80	334	66	0.356	0.237
Quezaltenango	14 51.0	-91-31.0	–	16:04:06.4			61	80	334	66	0.365	0.245
HONDURAS												
San Pedro Sula	15 27.0	-88-02.0	–	16:14:21.5			67	81	335	67	0.331	0.213
Tegucigalpa	14 06.0	-87-13.0	–	16:13:54.9			68	78	334	71	0.284	0.171
NICARAGUA												
Bluefields	12 00.0	-83-49.0	–	16:20:12.6			72	70	335	81	0.181	0.089
Granada	11 56.0	-85-58.0	–	16:13:12.2			68	72	334	77	0.210	0.110
Leon	12 25.0	-86-53.0	–	16:11:27.4			67	74	334	74	0.235	0.130
Managua	12 09.0	-86-17.0	–	16:12:41.3			68	73	334	76	0.220	0.118
PANAMA												
Colon	9 22.0	-79-54.0	–	16:29:16.9			76	53	335	100	0.059	0.017
David	8 26.0	-82-26.0	–	16:17:44.5			71	59	334	91	0.068	0.021
Panama City	8 58.0	-79-31.0	–	16:30:02.1			76	50	335	103	0.043	0.011
San Miguelito	11 24.0	-84-54.0	–	16:15:26.1			70	70	334	80	0.181	0.088
ANGUILLA												
The Valley	18 15.0	-63-05.0	–	18:03:41.1			63	273	348	263	0.206	0.107
ANTIGUA												
St. Johns	17 06.0	-61-51.0	–	18:08:33.9			60	276	349	262	0.172	0.082
THE BAHAMAS												
Nassau	25 05.0	-77-21.0	4	17:09:37.6			83	187	340	333	0.470	0.352
Freeport	26 30.0	-78-45.0	–	17:06:40.6			81	173	340	346	0.521	0.407
BARBADOS												
Bridgetown	13 06.0	-59-37.0	59	18:17:09.7			55	282	350	258	0.056	0.015
CUBA												
Havana	23 08.0	-82-22.0	26	16:48:01.0			80	120	337	35	0.463	0.345
Camaguey	21 23.0	-77-55.0	–	17:01:19.3			86	157	339	2	0.369	0.250
Cienfuegos	22 09.0	-80-27.0	–	16:52:58.3			83	126	338	31	0.416	0.296
Guantanamo	20 08.0	-75-12.0	–	17:10:35.3			86	233	340	287	0.310	0.194
Santiago de Cuba	20 01.0	-75-49.0	–	17:07:43.1			87	220	339	299	0.312	0.196
DOMINICAN REPUBLIC												
Santiago	19 27.0	-70-42.0	–	17:30:03.9			78	264	343	262	0.260	0.151
Santo Domingo	18 28.0	-69-54.0	19	17:32:43.4			76	269	343	258	0.227	0.124
MARTINIQUE												
Fort-de-France	14 36.0	-61-05.0	4	18:11:17.8			59	280	349	258	0.098	0.036
GUADELOUPE												
Basse-Terre	16 00.0	-61-44.0	574	18:08:50.8			60	278	349	260	0.139	0.060
HAITI												
Port-au-Prince	18 32.0	-72-20.0	40	17:21:13.1			82	266	341	258	0.243	0.136
JAMAICA												
Kingston	18 00.0	-76-48.0	36	16:59:54.2			89	110	338	55	0.263	0.153
PUERTO RICO												
San Juan	18 28.0	-66-07.0	4	17:50:28.5			69	271	346	261	0.215	0.114
Ponce	18 01.0	-66-37.0	–	17:47:52.5			70	272	345	259	0.203	0.104
ST. KITTS & NEVIS												
Basseterre	15 18.0	-62-43.0	–	18:04:22.4			62	279	348	257	0.117	0.047
MONSERRAT												
Plymouth	16 43.0	-62-12.0	–	18:07:01.5			61	276	349	261	0.160	0.074
US VIRGIN ISLANDS												
Charlotte Am., St..	18 21.0	-64-56.0	4	17:55:43.9			66	272	347	262	0.210	0.110
Christiansted, St..	17 45.0	-64-42.0	–	17:56:29.2			66	274	347	261	0.191	0.096

LOCAL CIRCUMSTANCES DURING THE ANNULAR SOLAR ECLIPSE OF 10 MAY 1994
FOR CENTRAL AMERICA AND CARIBBEAN

Location Name	First Contact U.T. h m s	Alt	P	V	Second Contact U.T. h m s	Alt	P	V	Third Contact U.T. h m s	Alt	P	V	Fourth Contact U.T. h m s	Alt	P	V
BELIZE																
Belize City	14:54:44.9	49	283	7									18:01:23.6	87	32	296
Belmopan	14:53:14.4	48	283	7									17:58:48.2	88	31	284
COSTA RICA																
Cartago	15:20:53.9	57	306	45									17:24:27.0	82	7	174
Limon	15:25:20.0	59	307	48									17:26:30.9	82	7	183
Puntarenas	15:15:50.6	55	304	42									17:24:26.4	82	9	169
San Jose	15:19:42.7	57	305	45									17:24:58.3	82	8	174
EL SALVADOR																
San Miguel	14:56:37.4	48	291	20									17:41:40.9	85	22	180
San Salvador	14:53:14.4	47	289	18									17:40:42.4	85	24	167
Santa Ana	14:52:03.0	46	288	16									17:41:30.0	85	25	165
GUATEMALA																
Antigua	14:48:38.7	44	286	13									17:41:39.9	85	27	154
Guatemala City	14:49:03.8	45	286	13									17:42:28.3	85	27	156
Mazatenango	14:46:36.2	43	285	12									17:39:43.2	84	28	148
Quezaltenango	14:46:29.1	43	284	11									17:41:14.1	84	28	148
HONDURAS																
San Pedro Sula	14:55:44.8	49	287	14									17:52:18.8	88	27	228
Tegucigalpa	14:59:10.0	50	291	20									17:47:14.9	86	23	210
NICARAGUA																
Bluefields	15:15:22.7	57	300	35									17:41:22.0	84	14	217
Granada	15:06:22.8	52	297	31									17:36:55.2	84	16	188
Leon	15:02:09.4	51	295	27									17:38:14.1	85	18	184
Managua	15:04:47.4	52	296	29									17:37:43.6	84	17	188
PANAMA																
Colon	15:49:20.0	67	316	65									17:21:43.3	82	359	189
David	15:36:30.4	62	314	60									17:11:06.4	80	359	157
Panama City	15:55:53.2	69	319	71									17:16:05.0	81	356	179
San Miguelito	15:11:52.2	55	300	35									17:35:10.0	84	13	192
ANGUILLA																
The Valley	16:45:19.0	81	309	223									19:18:42.8	45	31	309
ANTIGUA																
St. Johns	16:57:03.9	77	313	222									19:17:29.5	44	28	305
THE BAHAMAS																
Nassau	15:28:21.6	66	280	348									18:59:26.5	63	45	335
Freeport	15:24:02.1	64	277	343									18:58:37.5	64	48	342
BARBADOS																
Bridgetown	17:38:01.2	65	331	234									18:59:56.7	45	13	284
CUBA																
Havana	15:12:01.1	58	279	354									18:39:24.8	72	41	331
Camaguey	15:27:25.1	66	287	4									18:47:01.4	66	37	319
Cienfuegos	15:18:03.8	61	283	359									18:42:01.6	70	39	325
Guantanamo	15:39:45.5	72	292	12									18:50:35.9	63	33	313
Santiago de C..	15:37:13.5	70	292	12									18:48:16.7	64	33	312
DOMINICAN REPUBLIC																
Santiago	16:02:17.0	81	299	16									19:01:25.5	56	32	310
Santo Domingo	16:09:10.8	84	302	25									18:59:43.4	56	29	306
MARTINIQUE																
Fort-de-France	17:17:25.4	71	323	226									19:06:29.1	45	19	292
GUADELOUPE																
Basse-Terre	17:04:11.1	75	317	222									19:12:37.9	45	24	299
HAITI																
Port-au-Prince	15:56:23.9	78	299	24									18:52:36.6	60	29	305
JAMAICA																
Kingston	15:35:50.3	69	295	21									18:36:25.6	68	28	302
PUERTO RICO																
San Juan	16:29:03.4	88	306	235									19:10:58.0	50	30	308
Ponce	16:28:16.5	89	307	227									19:07:36.2	51	29	305
ST. KITTS & NEVIS																
Basseterre	17:04:06.6	76	319	221									19:05:37.6	47	21	295
MONSERRAT																
Plymouth	16:57:32.4	77	314	221									19:14:45.3	45	26	302
US VIRGIN ISLANDS																
Charlotte Am....	16:35:37.8	85	307	225									19:13:54.2	48	30	308
Christiansted..	16:39:31.2	84	309	220									19:12:05.3	48	29	305

Table 10a
CIRCUMSTANCES AT MAXIMUM ECLIPSE ON 10 MAY 1994
FOR THE UNITED STATES OF AMERICA

Location Name	Latitude ° ′	Longitude ° ′	Elev. m	U.T. h m s	Umbral Durat. m s	Path Width km	Sun Alt °	Sun Az. °	P °	V °	Eclipse Mag.	Eclipse Obs.
ALABAMA												
Anniston	33 39.0	-85-47.0	–	16:56:17.8			71	146	339	8	0.780	0.707
Birmingham	33 31.8	-86-48.6	203	16:53:21.8			71	141	339	12	0.787	0.715
Gadsden	34 00.6	-86-00.6	182	16:56:16.5			71	146	339	8	0.792	0.721
Huntsville	34 43.9	-86-35.2	210	16:55:55.7			70	146	339	9	0.816	0.750
Mobile	30 40.8	-88-06.6	2	16:44:53.1			71	128	338	23	0.727	0.643
Montgomery	32 21.6	-86-18.0	52	16:52:44.9			72	140	339	13	0.752	0.673
Tuscaloosa	33 12.0	-87-32.4	–	16:50:53.4			70	138	338	15	0.786	0.714
ALASKA												
Anchorage	61 12.0	-149-48.0	28	16:37:26.2			21	90	165	195	0.165	0.077
Fairbanks	64 50.0	-147-48.0	143	16:44:49.6			22	95	165	192	0.147	0.065
Juneau	58 18.2	-134-24.5	4	16:36:30.5			28	103	164	196	0.276	0.164
ARIZONA												
Flagstaff	35 12.6	-111-37.2	2264	16:09:45.4			44	99	158	216	0.842	0.780
Glendale	33 30.0	-112-15.0	–	16:05:39.5			43	97	158	219	0.871	0.813
Mesa	33 25.0	-111-50.0	–	16:05:59.3			43	97	158	219	0.877	0.821
Phoenix	33 30.0	-112-04.8	366	16:05:51.3			43	97	158	219	0.872	0.816
Scottsdale	33 30.0	-111-53.0	–	16:06:05.5			43	97	158	218	0.875	0.818
Tempe	33 24.0	-111-54.0	–	16:05:52.5			43	97	158	219	0.877	0.820
Tucson	32 13.2	-110-55.2	784	16:04:44.1			44	96	158	220	0.912	0.861
Yuma	32 42.0	-114-37.8	52	16:01:26.5			40	94	158	220	0.860	0.801
ARKANSAS												
Fort Smith	35 22.8	-94-24.0	144	16:38:37.3			62	124	338	24	0.911	0.861
Little Rock	34 44.4	-92-19.2	94	16:42:00.5			65	127	338	22	0.874	0.819
N Little Rock	34 46.0	-92-13.0	–	16:42:17.5			65	127	338	22	0.873	0.818
Pine Bluff	34 13.2	-92-01.2	–	16:41:44.4			65	127	338	22	0.858	0.800
CALIFORNIA												
Alameda	37 46.0	-122-15.0	–	16:04:14.8			34	93	160	216	0.681	0.588
Alhambra	34 05.0	-118-08.0	–	16:00:39.6			37	93	159	219	0.794	0.722
Anaheim	33 50.0	-117-55.0	–	16:00:22.8			37	93	159	219	0.801	0.730
Bakersfield	35 23.0	-119 00.0	131	16:02:22.5			36	93	159	218	0.759	0.680
Baldwin Park	34 05.0	-117-58.0	–	16:00:48.8			37	93	159	219	0.796	0.724
Bellflower	33 53.0	-118-08.0	–	16:00:16.6			37	93	159	219	0.798	0.726
Berkeley	37 52.0	-122-17.0	13	16:04:24.8			34	93	160	216	0.679	0.585
Buena Park	33 52.0	-118 00.0	–	16:00:22.0			37	93	159	219	0.800	0.729
Burbank	34 11.0	-118-19.0	–	16:00:41.1			37	93	159	219	0.790	0.717
Carson	33 49.0	-118-16.0	–	16:00:01.7			37	92	159	219	0.798	0.726
Cerritos	33 52.0	-118-05.0	–	16:00:17.4			37	93	159	219	0.799	0.727
Chula Vista	32 38.0	-117-05.0	–	15:58:51.6			37	92	158	221	0.834	0.770
Compton	33 54.0	-118-14.0	–	16:00:13.1			37	93	159	219	0.796	0.725
Concord	37 58.0	-122-02.0	–	16:04:48.1			34	94	160	215	0.680	0.586
Cosa Mesa	33 39.0	-118-54.0	–	15:59:08.7			36	92	159	220	0.794	0.722
Daly City	37 43.0	-122-31.0	–	16:03:56.6			34	93	160	216	0.679	0.586
Downey	33 56.0	-118-08.0	–	16:00:22.4			37	93	159	219	0.797	0.725
El Cajon	32 48.0	-116-58.0	–	15:59:17.4			38	92	158	220	0.832	0.767
El Monte	34 04.0	-118-02.0	–	16:00:43.2			37	93	159	219	0.795	0.723
Escondido	33 07.0	-117 00.0	–	15:59:51.9			38	93	158	220	0.825	0.759
Eureka	40 45.0	-124-10.0	–	16:08:32.9			33	95	160	213	0.609	0.505
Fairfield	38 14.0	-122-02.0	–	16:05:18.9			34	94	160	215	0.675	0.580
Fountain Valley	33 42.0	-117-57.0	–	16:00:05.6			37	93	159	220	0.803	0.733
Fremont	37 33.0	-122 00.0	–	16:04:01.6			34	93	159	216	0.688	0.596
Fresno	36 46.2	-119-46.8	94	16:04:21.8			36	94	159	216	0.725	0.639
Fullerton	33 53.0	-117-56.0	–	16:00:27.6			37	93	159	219	0.800	0.729
Garden Grove	33 47.0	-117-56.0	–	16:00:16.1			37	93	159	219	0.802	0.731
Glendale	34 09.0	-118-15.0	–	16:00:40.9			37	93	159	219	0.791	0.719
Hawthorne	33 55.0	-118-22.0	–	16:00:07.8			37	92	159	219	0.795	0.723
Hayward	37 40.0	-122-06.0	–	16:04:10.3			34	93	160	216	0.685	0.592
Huntington Beach	33 39.0	-118 00.0	–	15:59:57.1			37	92	159	220	0.804	0.734
Inglewood	33 57.0	-118-22.0	–	16:00:11.6			37	92	159	219	0.794	0.722
Irvine	33 40.0	-117-45.0	–	16:00:12.9			37	93	159	220	0.806	0.737
Lakewood	33 50.0	-118-09.0	–	16:00:10.0			37	93	159	219	0.799	0.727
La Mesa	32 46.0	-117-01.0	–	15:59:10.7			38	92	158	220	0.832	0.768
Long Beach	33 46.0	-118-12.0	–	15:59:59.6			37	92	159	220	0.799	0.728
Los Angeles	34 04.8	-118-22.2	32	16:00:26.3			37	93	159	219	0.791	0.719
Modesto	37 39.0	-121 00.0	–	16:05:01.6			35	94	159	216	0.696	0.605

46

LOCAL CIRCUMSTANCES DURING THE ANNULAR SOLAR ECLIPSE OF 10 MAY 1994
FOR THE UNITED STATES OF AMERICA

Location Name	First Contact				Second Contact				Third Contact				Fourth Contact			
	U.T. h m s	Alt °	P °	V °	U.T. h m s	Alt °	P °	V °	U.T. h m s	Alt °	P °	V °	U.T. h m s	Alt °	P °	V °
ALABAMA																
Anniston	15:13:23.4	54	261	318									18:50:10.5	68	62	21
Birmingham	15:11:13.5	52	260	318									18:47:16.6	69	62	23
Gadsden	15:13:28.6	53	260	317									18:49:56.7	67	63	22
Huntsville	15:13:29.3	53	259	314									18:49:07.0	67	64	26
Mobile	15:04:39.7	51	263	326									18:39:45.6	73	58	18
Montgomery	15:10:31.2	53	262	322									18:47:15.4	69	60	18
Tuscaloosa	15:09:21.7	52	260	319									18:44:54.1	70	62	24
ALASKA																
Anchorage	15:58:56.8	16	200	230									17:16:31.9	25	131	161
Fairbanks	16:07:51.4	18	198	225									17:22:00.5	26	134	160
Juneau	15:45:18.4	22	209	243									17:29:32.7	35	120	150
ARIZONA																
Flagstaff	14:48:04.8	27	242	301									17:44:55.2	62	77	123
Glendale	14:44:25.9	26	244	305									17:40:55.7	62	75	126
Mesa	14:44:25.2	26	244	305									17:41:44.0	62	74	125
Phoenix	14:44:29.3	26	244	305									17:41:18.4	62	75	126
Scottsdale	14:44:33.9	26	244	305									17:41:45.1	62	74	125
Tempe	14:44:21.7	26	244	305									17:41:33.4	62	74	125
Tucson	14:42:28.0	26	246	308									17:41:52.2	64	72	124
Yuma	14:42:08.7	23	243	305									17:34:27.6	59	75	131
ARKANSAS																
Fort Smith	15:02:19.7	44	253	310									18:28:48.0	72	68	58
Little Rock	15:04:01.3	46	255	313									18:33:40.9	72	66	47
N Little Rock	15:04:13.3	46	255	313									18:33:59.6	72	66	47
Pine Bluff	15:03:33.1	47	256	314									18:33:52.6	72	65	44
CALIFORNIA																
Alameda	14:51:43.3	20	234	289									17:26:46.8	50	87	139
Alhambra	14:44:12.9	21	240	300									17:29:24.7	55	79	135
Anaheim	14:43:44.5	21	240	300									17:29:27.8	55	79	135
Bakersfield	14:46:46.3	21	238	296									17:29:36.7	54	82	136
Baldwin Park	14:44:14.1	21	240	300									17:29:44.5	55	79	135
Bellflower	14:43:48.9	21	240	300									17:29:06.6	55	79	135
Berkeley	14:51:56.4	20	234	289									17:26:51.0	50	87	139
Buena Park	14:43:47.8	21	240	300									17:29:20.9	55	79	135
Burbank	14:44:23.6	21	240	299									17:29:11.9	55	80	135
Carson	14:43:40.0	21	240	300									17:28:44.8	55	79	135
Cerritos	14:43:47.2	21	240	300									17:29:11.0	55	79	135
Chula Vista	14:41:30.5	21	242	303									17:29:16.6	56	77	134
Compton	14:43:50.2	21	240	300									17:28:56.3	55	79	135
Concord	14:52:09.3	20	234	289									17:27:26.2	50	87	139
Cosa Mesa	14:43:16.2	20	240	300									17:27:15.5	54	79	136
Daly City	14:51:37.1	20	234	289									17:26:14.0	50	87	139
Downey	14:43:54.9	21	240	300									17:29:11.1	55	79	135
El Cajon	14:41:51.0	21	242	303									17:29:46.4	56	77	134
El Monte	14:44:11.6	21	240	300									17:29:35.0	55	79	135
Escondido	14:42:27.6	21	242	302									17:30:12.1	56	77	134
Eureka	14:58:29.1	20	230	282									17:27:08.9	48	92	141
Fairfield	14:52:44.4	20	233	289									17:27:47.1	50	88	139
Fountain Vall...	14:43:28.3	21	240	301									17:29:11.8	55	79	135
Fremont	14:51:14.8	20	234	290									17:26:56.8	50	87	139
Fresno	14:49:37.7	21	236	293									17:30:02.7	53	84	136
Fullerton	14:43:50.3	21	240	300									17:29:30.4	55	79	135
Garden Grove	14:43:38.4	21	240	300									17:29:21.3	55	79	135
Glendale	14:44:20.1	21	240	299									17:29:16.8	55	80	135
Hawthorne	14:43:51.2	21	240	300									17:28:42.1	55	79	135
Hayward	14:51:30.1	20	234	290									17:26:55.2	50	87	139
Huntington Be..	14:43:22.0	21	240	301									17:29:01.3	55	79	135
Inglewood	14:43:55.2	21	240	300									17:28:45.1	55	79	135
Irvine	14:43:25.9	21	241	301									17:29:32.6	55	79	135
Lakewood	14:43:42.8	21	240	300									17:29:00.1	55	79	135
La Mesa	14:41:46.6	21	242	303									17:29:37.2	56	77	134
Long Beach	14:43:34.5	21	240	300									17:28:48.2	55	79	135
Los Angeles	14:44:10.8	21	240	299									17:28:56.3	55	80	135
Modesto	14:51:28.3	21	235	290									17:28:55.4	51	86	137

Table 10a
CIRCUMSTANCES AT MAXIMUM ECLIPSE ON 10 MAY 1994
FOR THE UNITED STATES OF AMERICA

Location Name	Latitude ° ′	Longitude ° ′	Elev. m	U.T. h m s	Umbral Durat. m s	Path Width km	Sun Alt °	Sun Az. °	P °	V °	Eclipse Mag.	Eclipse Obs.
CALIFORNIA												
Montebello	34 01.0	-118-06.0	—	16:00:33.8			37	93	159	219	0.796	0.724
Monterey Park	34 04.0	-118-08.0	—	16:00:37.7			37	93	159	219	0.794	0.722
Mountain View	37 25.0	-122-07.0	—	16:03:40.7			34	93	159	216	0.689	0.597
Napa	38 20.0	-122-17.0	—	16:05:18.6			34	94	160	215	0.670	0.575
Newport Beach	33 36.0	-117-55.0	—	15:59:56.0			37	92	159	220	0.806	0.736
Norwalk	33 54.0	-118-05.0	—	16:00:21.3			37	93	159	219	0.798	0.727
Oakland	37 48.0	-122-16.0	8	16:04:17.9			34	93	160	216	0.680	0.587
Oceanside	33 11.0	-117-22.0	—	15:59:38.7			37	92	158	220	0.820	0.753
Ontario	34 04.0	-117-39.0	—	16:01:04.5			37	93	159	219	0.799	0.728
Orange	33 48.0	-117-51.0	—	16:00:22.6			37	93	159	219	0.803	0.732
Oxnard	34 08.0	-119-12.0	—	15:59:48.3			36	92	159	219	0.781	0.707
Palo Alto	37 27.0	-122-09.0	—	16:03:43.0			34	93	160	216	0.688	0.596
Pasadena	34 09.0	-118-09.0	272	16:00:46.2			37	93	159	219	0.792	0.720
Pico Rivera	34 01.0	-118-05.0	—	16:00:34.7			37	93	159	219	0.796	0.724
Pomona	34 04.0	-117-45.0	—	16:00:58.9			37	93	159	219	0.798	0.727
Rancho Cucamonga	34 05.0	-117-35.0	—	16:01:10.2			37	93	159	219	0.800	0.729
Redondo Beach	33 50.0	-118-23.0	—	15:59:57.3			36	92	159	219	0.796	0.724
Redwood City	37 29.0	-122-13.0	—	16:03:43.7			34	93	160	216	0.687	0.594
Richmond	37 56.0	-122-21.0	—	16:04:29.3			34	93	160	216	0.677	0.583
Riverside	33 59.0	-117-21.0	—	16:01:11.8			38	93	159	219	0.804	0.734
Sacramento	38 35.0	-121-30.0	10	16:06:25.2			35	95	160	215	0.674	0.579
Salinas	36 41.0	-121-40.0	—	16:02:37.7			34	93	159	217	0.707	0.618
San Bernardino	34 07.0	-117-19.0	354	16:01:28.9			38	93	159	219	0.802	0.731
San Buenaventura	34 18.0	-119-18.0	—	16:00:02.2			36	92	159	219	0.777	0.702
San Diego	32 45.0	-117-08.4	7	15:59:01.8			37	92	158	221	0.831	0.766
San Francisco	37 45.6	-122-26.4	21	16:04:05.2			34	93	160	216	0.679	0.586
San Jose	37 20.0	-121-54.0	30	16:03:41.4			34	93	159	216	0.693	0.601
San Leandro	37 43.0	-122-10.0	—	16:04:12.9			34	93	160	216	0.683	0.590
San Mateo	37 34.0	-122-20.0	—	16:03:47.9			34	93	160	216	0.684	0.591
Santa Ana	33 41.0	-117-57.0	—	16:00:03.7			37	93	159	220	0.804	0.734
Santa Barbara	34 26.0	-119-43.0	33	15:59:55.9			35	92	159	219	0.770	0.693
Santa Clara	37 21.0	-121-56.0	—	16:03:41.7			34	93	159	216	0.692	0.601
Santa Monica	34 01.0	-118-29.0	—	16:00:12.9			36	92	159	219	0.791	0.719
Santa Rosa	38 27.0	-122-42.0	—	16:05:12.7			34	94	160	215	0.664	0.568
Simi Valley	34 16.0	-118-47.0	—	16:00:25.6			36	92	159	219	0.783	0.709
South Gate	33 57.0	-118-13.0	—	16:00:19.7			37	93	159	219	0.796	0.724
Stockton	37 57.5	-121-17.3	7	16:05:23.1			35	94	159	215	0.687	0.595
Sunnyvale	37 23.0	-122-02.0	—	16:03:40.8			34	93	159	216	0.690	0.599
Thousand Oaks	34 10.0	-118-50.0	—	16:00:11.4			36	92	159	219	0.785	0.711
Torrance	33 50.0	-118-20.0	—	15:59:60.0			37	92	159	219	0.797	0.725
Vallejo	38 06.0	-122-15.0	—	16:04:53.3			34	94	160	215	0.675	0.581
Visalia	36 20.0	-119-18.0	—	16:03:56.5			36	94	159	217	0.738	0.655
Walnut Creek	37 54.0	-122-04.0	—	16:04:38.8			34	94	160	216	0.681	0.587
West Covina	34 04.0	-117-55.0	—	16:00:49.6			37	93	159	219	0.797	0.725
Westminster	33 45.0	-117-59.0	—	16:00:09.5			37	93	159	219	0.802	0.732
Whittier	33 58.0	-118-02.0	—	16:00:31.7			37	93	159	219	0.797	0.726
COLORADO												
Arvada	39 48.0	-105-05.0	—	16:27:44.8			51	114	159	207	0.818	0.752
Aurora	39 43.0	-104-49.0	—	16:27:59.4			51	114	159	206	0.823	0.757
Boulder	40 00.2	-105-15.7	—	16:27:52.1			51	114	159	206	0.812	0.745
Colorado Springs	38 49.0	-104-48.0	1932	16:26:15.9			51	112	159	208	0.842	0.780
Denver	39 43.2	-104-58.8	1732	16:27:44.2			51	114	159	207	0.821	0.755
Durango	37 15.0	-107-55.0	—	16:18:38.8			48	106	158	212	0.841	0.779
Fort Collins	40 36.0	-105-04.0	—	16:29:18.3			51	115	159	205	0.802	0.732
Grand Junction	39 04.2	-108-33.0	1506	16:21:20.6			48	108	159	210	0.797	0.726
Greeley	40 25.0	-104-41.0	—	16:29:32.1			51	115	159	205	0.810	0.741
Lakewood	39 44.0	-105-06.0	—	16:27:35.6			51	113	159	207	0.820	0.753
Pueblo	38 17.4	-104-38.4	1539	16:25:29.4			52	112	158	209	0.855	0.795
Westminster	39 50.0	-105-02.0	—	16:27:53.1			51	114	159	206	0.818	0.752

Table 10b
LOCAL CIRCUMSTANCES DURING THE ANNULAR SOLAR ECLIPSE OF 10 MAY 1994
FOR THE UNITED STATES OF AMERICA

Location Name	First Contact U.T. h m s	Alt °	P °	V °	Second Contact U.T. h m s	Alt °	P °	V °	Third Contact U.T. h m s	Alt °	P °	V °	Fourth Contact U.T. h m s	Alt °	P °	V °
CALIFORNIA																
Montebello	14:44:05.1	21	240	300									17:29:22.6	55	79	135
Monterey Park	14:44:10.9	21	240	300									17:29:23.2	55	79	135
Mountain View	14:50:57.6	20	234	290									17:26:33.5	50	87	139
Napa	14:52:57.7	20	233	288									17:27:27.6	50	88	139
Newport Beach	14:43:16.7	21	241	301									17:29:06.6	55	79	135
Norwalk	14:43:51.2	21	240	300									17:29:14.1	55	79	135
Oakland	14:51:47.6	20	234	289									17:26:47.6	50	87	139
Oceanside	14:42:32.0	21	241	302									17:29:34.0	56	78	134
Ontario	14:44:14.6	21	240	300									17:30:20.8	55	79	134
Orange	14:43:41.0	21	240	300									17:29:32.8	55	79	135
Oxnard	14:44:12.5	20	239	299									17:27:24.2	54	80	136
Palo Alto	14:51:01.9	20	234	290									17:26:32.5	50	87	139
Pasadena	14:44:20.6	21	240	299									17:29:28.5	55	80	135
Pico Rivera	14:44:05.2	21	240	300									17:29:24.6	55	79	135
Pomona	14:44:13.8	21	240	300									17:30:08.8	55	79	135
Rancho Cucamo...	14:44:17.1	22	240	300									17:30:30.3	56	79	134
Redondo Beach	14:43:41.2	21	240	300									17:28:32.6	55	79	135
Redwood City	14:51:06.3	20	234	290									17:26:27.9	50	87	139
Richmond	14:52:05.2	20	234	289									17:26:49.1	50	87	139
Riverside	14:44:07.1	22	240	300									17:30:49.4	56	79	134
Sacramento	14:53:30.7	21	233	288									17:29:12.9	51	88	138
Salinas	14:49:23.1	20	235	292									17:26:23.4	51	85	139
San Bernardino	14:44:23.3	22	240	300									17:31:05.2	56	79	134
San Buenavent...	14:44:32.1	20	239	298									17:27:27.4	53	81	137
San Diego	14:41:43.5	21	242	303									17:29:20.8	56	77	134
San Francisco	14:51:42.6	20	234	289									17:26:25.6	50	87	139
San Jose	14:50:46.7	20	234	290									17:26:50.4	50	86	139
San Leandro	14:51:36.7	20	234	290									17:26:51.9	50	87	139
San Mateo	14:51:17.3	20	234	290									17:26:21.9	50	87	139
Santa Ana	14:43:26.3	21	240	301									17:29:10.2	55	79	135
Santa Barbara	14:44:46.5	20	239	298									17:26:51.3	53	81	137
Santa Clara	14:50:48.8	20	234	290									17:26:48.1	50	86	139
Santa Monica	14:44:02.5	21	240	300									17:28:37.3	54	80	136
Santa Rosa	14:53:13.7	20	233	288									17:26:51.6	49	88	139
Simi Valley	14:44:30.8	21	239	299									17:28:24.5	54	80	136
South Gate	14:43:56.3	21	240	300									17:29:02.8	55	79	135
Stockton	14:52:08.3	21	234	290									17:28:47.6	51	87	138
Sunnyvale	14:50:53.2	20	234	290									17:26:39.9	50	86	139
Thousand Oaks	14:44:18.4	21	239	299									17:28:09.8	54	80	136
Torrance	14:43:41.5	21	240	300									17:28:38.5	55	79	135
Vallejo	14:52:26.9	20	234	289									17:27:13.0	50	88	139
Visalia	14:48:44.1	21	237	294									17:30:22.2	53	83	136
Walnut Creek	14:52:00.6	20	234	289									17:27:17.3	50	87	139
West Covina	14:44:12.5	21	240	300									17:29:48.9	55	79	135
Westminster	14:43:34.0	21	240	300									17:29:12.3	55	79	135
Whittier	14:43:59.6	21	240	300									17:29:26.0	55	79	135
COLORADO																
Arvada	15:01:19.1	35	241	295									18:06:07.4	65	80	104
Aurora	15:01:18.4	35	241	295									18:06:41.0	66	79	103
Boulder	15:01:39.6	35	241	294									18:05:53.9	65	80	104
Colorado Spri...	14:59:18.8	35	242	297									18:05:46.2	66	78	103
Denver	15:01:11.6	35	241	295									18:06:17.2	65	79	103
Durango	14:54:04.0	31	242	299									17:56:22.5	65	77	113
Fort Collins	15:03:07.8	35	240	293									18:06:56.9	65	81	103
Grand Junction	14:57:42.1	32	240	295									17:57:04.8	63	80	114
Greeley	15:02:57.9	36	241	293									18:07:41.6	65	80	102
Lakewood	15:01:09.6	35	241	295									18:06:00.9	65	79	104
Pueblo	14:58:16.6	35	243	298									18:05:34.7	67	77	102
Westminster	15:01:25.5	35	241	295									18:06:16.7	65	80	103

CIRCUMSTANCES AT MAXIMUM ECLIPSE ON 10 MAY 1994
FOR THE UNITED STATES OF AMERICA

Location Name	Latitude ° ′	Longitude ° ′	Elev. m	U.T. h m s	Umbral Durat. m s	Path Width km	Sun Alt °	Sun Az. °	P °	V °	Eclipse Mag.	Eclipse Obs.
CONNECTICUT												
Bridgeport	41 11.4	-73-11.4	3	17:37:34.0			64	207	345	323	0.901	0.850
Bristol	41 40.0	-72-55.0	–	17:38:24.1			64	208	345	323	0.913	0.864
Danbury	41 23.0	-73-27.0	–	17:37:00.3			64	206	345	324	0.908	0.858
East Hartford	41 45.0	-72-35.0	–	17:39:13.9			63	209	345	322	0.914	0.865
Fairfield	41 08.0	-73-22.0	–	17:37:06.7			64	207	345	323	0.900	0.849
Greenwich	41 01.0	-73-37.0	–	17:36:26.9			65	206	345	324	0.898	0.847
Hamden	41 20.0	-72-55.0	–	17:38:17.3			64	208	345	323	0.904	0.854
Hartford	41 45.6	-72-41.4	13	17:38:58.7			64	209	345	323	0.915	0.866
Manchester	41 45.0	-72-30.0	–	17:39:25.9			63	209	345	322	0.914	0.865
Meriden	41 30.0	-72-50.0	62	17:38:32.9			64	208	345	323	0.908	0.858
Milford	41 15.0	-73-05.0	–	17:37:51.1			64	208	345	323	0.902	0.852
New Britain	41 40.0	-72-45.0	66	17:38:48.3			64	209	345	323	0.913	0.863
New Haven	41 18.6	-72-55.8	13	17:38:14.8			64	208	345	323	0.903	0.853
Norwalk	41 06.0	-73-25.0	–	17:36:58.5			65	207	345	324	0.900	0.849
Stamford	41 03.0	-73-32.0	11	17:36:40.1			65	206	345	324	0.899	0.848
Stratford	41 10.0	-73-05.0	–	17:37:49.2			64	208	345	323	0.900	0.849
Waterbury	41 30.0	-73 00.0	85	17:38:08.6			64	208	345	323	0.909	0.859
West Hartford	41 45.0	-72-45.0	–	17:38:49.9			64	208	345	323	0.915	0.866
West Haven	41 16.0	-72-57.0	–	17:38:11.0			64	208	345	323	0.902	0.852
DELAWARE												
Dover	39 09.6	-75-31.8	–	17:30:35.3			67	200	344	327	0.855	0.797
Wilmington	39 45.0	-75-33.0	44	17:30:54.6			67	200	344	327	0.872	0.816
DISTRICT OF COLUMBIA												
Washington	38 52.8	-77-01.2	5	17:26:28.1			68	194	343	331	0.855	0.797
FLORIDA												
Boca Raton	26 21.0	-80-05.0	–	17:01:43.9			81	157	339	0	0.529	0.416
Clearwater	27 43.0	-82-45.0	–	16:55:10.4			78	142	338	14	0.592	0.487
Daytona Beach	29 11.0	-81-02.0	2	17:03:08.2			78	160	339	358	0.616	0.514
Fort Lauderdale	26 07.0	-80-09.0	–	17:01:06.6			81	156	339	2	0.523	0.409
Gainesville	29 39.6	-82-19.8	57	16:59:47.3			77	152	339	4	0.641	0.543
Hialeah	25 49.0	-80-18.0	–	17:00:04.7			81	153	339	5	0.516	0.402
Hollywood	26 00.0	-80-11.0	–	17:00:47.8			81	155	339	3	0.520	0.406
Jacksonville	30 19.2	-81-39.0	7	17:02:57.8			77	159	339	359	0.653	0.556
Largo	27 54.0	-82-47.0	–	16:55:23.3			77	142	338	13	0.598	0.493
Miami	25 46.8	-80-13.2	2	17:00:17.8			81	153	339	4	0.515	0.400
Orlando	28 32.4	-81-22.8	23	17:00:58.8			78	155	339	2	0.602	0.497
Pensacola	30 25.0	-87-13.0	5	16:46:48.6			72	130	338	22	0.710	0.624
Pompano Beach	26 12.0	-80-07.0	–	17:01:21.9			81	156	339	1	0.525	0.412
St. Petersburg	27 47.0	-82-38.0	7	16:55:39.8			78	143	338	13	0.593	0.487
Sarasota	27 20.0	-82-32.0	7	16:55:11.8			78	142	338	14	0.580	0.472
Tallahassee	30 26.4	-84-17.4	–	16:55:09.4			75	143	339	12	0.681	0.589
Tampa	27 57.6	-82-28.2	–	16:56:29.5			78	145	338	11	0.596	0.491
West Palm Beach	26 43.0	-80-03.2	–	17:02:26.5			80	159	339	359	0.539	0.427
GEORGIA												
Albany	31 34.8	-84-09.6	–	16:57:26.5			74	148	339	8	0.710	0.624
Atlanta	33 45.6	-84-24.6	331	17:00:12.8			72	153	339	3	0.771	0.696
Augusta	33 28.2	-81-59.4	47	17:06:38.0			74	165	340	353	0.742	0.661
Columbus	32 28.8	-84-57.0	87	16:56:40.3			73	147	339	8	0.742	0.661
Macon	32 49.8	-83-39.6	110	17:00:52.5			73	154	339	2	0.739	0.658
Savannah	32 03.0	-81-05.4	7	17:07:16.8			75	167	340	351	0.696	0.606
HAWAII												
Hilo	19 44.0	-155-01.0	13	15:47 Rise			0	71	–	–	0.659	0.560
Honolulu	21 18.6	-157-50.4	7	15:56 Rise			0	71	–	–	0.549	0.435
IDAHO												
Boise	43 36.6	-116-13.2	931	16:21:07.1			41	105	160	207	0.633	0.532
Coeur D'Alene	47 40.8	-116-46.2	–	16:28:26.8			40	110	160	203	0.557	0.446
Lewiston	46 24.0	-116-59.0	–	16:25:46.4			40	108	160	204	0.577	0.468
Pocatello	42 52.8	-112-27.0	1463	16:23:50.5			44	109	160	207	0.683	0.590
Twin Falls	42 33.0	-114-29.0	–	16:20:54.0			42	106	160	208	0.669	0.574
ILLINOIS												
Arlington Heights	42 05.0	-87-59.0	–	17:03:28.8			64	155	161	180	0.928	0.879
Aurora	41 45.0	-88-18.0	–	17:02:20.4			64	154	161	181	0.934	0.884
Bloomington	40 29.0	-89 00.0	262	16:59:00.3	5 36.1	232	64	149	160	184	0.943	0.889
Champaign	40 06.6	-88-15.0	243	17:00:08.1	6 8.7	232	65	151	340	3	0.943	0.889
Chicago	41 51.0	-87-40.8	199	17:03:49.6			64	156	161	180	0.937	0.886
Cicero	41 50.0	-87-46.0	–	17:03:36.9			64	156	161	180	0.936	0.886
Decatur	39 50.0	-88-59.0	224	16:58:04.7	6 7.1	233	65	148	340	5	0.943	0.889

50

LOCAL CIRCUMSTANCES DURING THE ANNULAR SOLAR ECLIPSE OF 10 MAY 1994
FOR THE UNITED STATES OF AMERICA

Location Name	First Contact U.T. h m s	Alt °	P °	V °	Second Contact U.T. h m s	Alt °	P °	V °	Third Contact U.T. h m s	Alt °	P °	V °	Fourth Contact U.T. h m s	Alt °	P °	V °
CONNECTICUT																
Bridgeport	15:50:54.8	63	258	283									19:21:01.0	50	75	29
Bristol	15:52:04.6	63	257	281									19:21:16.1	50	76	30
Danbury	15:50:32.7	63	257	283									19:20:26.0	50	75	30
East Hartford	15:52:55.4	63	257	280									19:21:49.7	49	76	30
Fairfield	15:50:26.7	63	258	284									19:20:42.8	50	75	29
Greenwich	15:49:44.5	63	258	284									19:20:17.9	50	75	28
Hamden	15:51:41.9	63	258	282									19:21:27.0	50	75	29
Hartford	15:52:41.6	63	257	281									19:21:37.7	49	76	30
Manchester	15:53:06.8	63	257	280									19:21:58.7	49	76	30
Meriden	15:52:04.5	63	258	282									19:21:30.9	50	76	30
Milford	15:51:13.4	63	258	283									19:21:11.0	50	75	29
New Britain	15:52:27.2	63	257	281									19:21:34.4	49	76	30
New Haven	15:51:38.5	63	258	282									19:21:26.2	50	75	29
Norwalk	15:50:17.6	63	258	284									19:20:38.2	50	75	29
Stamford	15:49:58.2	63	258	284									19:20:26.4	50	75	29
Stratford	15:51:07.8	63	258	283									19:21:13.6	50	75	29
Waterbury	15:51:41.8	63	258	282									19:21:12.5	50	76	30
West Hartford	15:52:32.8	63	257	281									19:21:31.5	49	76	30
West Haven	15:51:32.8	63	258	282									19:21:25.3	50	75	29
DELAWARE																
Dover	15:43:09.7	63	260	293									19:17:06.8	53	71	24
Wilmington	15:43:49.7	63	259	291									19:16:57.1	53	72	26
DISTRICT OF COLUMBIA																
Washington	15:39:24.2	62	259	296									19:14:01.1	55	71	24
FLORIDA																
Boca Raton	15:19:55.6	62	276	343									18:54:30.8	66	48	343
Clearwater	15:13:22.9	58	271	338									18:49:43.1	69	51	352
Daytona Beach	15:19:03.8	60	271	333									18:57:12.2	65	53	355
Fort Lauderda...	15:19:37.7	62	276	344									18:53:48.3	66	47	342
Gainesville	15:16:09.7	58	269	331									18:54:22.4	67	55	359
Hialeah	15:19:03.6	61	276	345									18:52:41.4	67	47	341
Hollywood	15:19:28.9	62	276	344									18:53:26.8	66	47	341
Jacksonville	15:18:29.9	59	268	329									18:57:14.7	65	56	0
Largo	15:13:26.7	58	271	337									18:49:59.1	68	51	353
Miami	15:19:17.2	62	277	345									18:52:50.2	66	47	340
Orlando	15:17:37.6	60	271	335									18:55:08.9	66	52	353
Pensacola	15:06:03.8	52	264	327									18:41:51.6	72	57	14
Pompano Beach	15:19:45.8	62	276	344									18:54:05.2	66	47	342
St. Petersburg	15:13:44.2	58	271	337									18:50:11.1	68	51	352
Sarasota	15:13:38.9	58	272	339									18:49:33.9	69	50	350
Tallahassee	15:12:20.0	56	266	328									18:50:07.8	68	56	6
Tampa	15:14:18.1	58	271	337									18:50:59.7	68	51	352
West Palm Bea...	15:20:10.9	62	275	342									18:55:23.0	65	48	344
GEORGIA																
Albany	15:13:58.9	56	265	324									18:52:06.8	67	58	9
Atlanta	15:16:19.1	55	261	317									18:53:57.4	66	62	18
Augusta	15:21:11.7	58	264	318									19:00:01.5	63	62	12
Columbus	15:13:26.4	55	263	321									18:51:05.9	68	60	15
Macon	15:16:39.5	56	263	320									18:54:58.9	66	61	13
Savannah	15:21:39.6	60	266	323									19:00:56.5	63	59	6
HAWAII																
Hilo	–												16:34:47.4	10	85	157
Honolulu	–												16:35:55.6	8	88	159
IDAHO																
Boise	15:05:43.7	27	231	281									17:45:16.8	54	91	128
Coeur D'Alene	15:15:48.0	28	227	272									17:47:55.3	52	97	129
Lewiston	15:12:30.4	28	228	274									17:46:25.6	52	95	129
Pocatello	15:04:55.9	30	234	284									17:52:22.3	58	88	121
Twin Falls	15:03:33.8	28	233	284									17:47:45.4	56	89	126
ILLINOIS																
Arlington Hei...	15:24:18.2	51	248	292									18:49:17.2	62	77	52
Aurora	15:23:10.5	51	249	293									18:48:29.9	63	77	52
Bloomington	15:19:39.0	50	250	297	16:56:10.2	64	227	252	17:01:46.3	65	96	120	18:46:31.0	64	75	50
Champaign	15:20:03.8	51	251	298	16:57:04.5	65	259	284	17:03:13.2	65	64	86	18:48:13.3	64	74	47
Chicago	15:24:17.8	51	249	293									18:49:58.4	62	77	51
Cicero	15:24:08.0	51	249	293									18:49:46.0	62	77	51
Decatur	15:18:25.0	50	251	299	16:55:02.1	65	260	287	17:01:09.2	65	63	87	18:46:19.3	65	74	48

Table 10a
CIRCUMSTANCES AT MAXIMUM ECLIPSE ON 10 MAY 1994
FOR THE UNITED STATES OF AMERICA

Location Name	Latitude ° ′	Longitude ° ′	Elev. m	U.T. h m s	Umbral Durat. m s	Path Width km	Sun Alt °	Sun Az. °	P °	V °	Eclipse Mag.	Eclipse Obs.
ILLINOIS												
Des Plaines	42 02.0	-87-54.0	—	17:03:35.6			64	155	161	180	0.930	0.881
East St. Louis	38 38.0	-90-10.0	—	16:53:34.1	3 31.7	234	65	142	339	10	0.943	0.889
Elgin	42 03.0	-88-16.0	—	17:02:49.3			64	154	161	181	0.927	0.878
Evanston	42 02.0	-87-41.0	—	17:04:03.9			64	156	161	179	0.932	0.882
Joliet	41 37.0	-88-05.0	—	17:02:37.7			64	154	161	181	0.939	0.888
Mount Prospect	42 03.0	-87-56.0	—	17:03:32.6			64	155	161	180	0.930	0.880
Oak Lawn	41 43.0	-87-45.0	—	17:03:29.7			64	155	161	180	0.939	0.888
Oak Park	41 53.0	-87-48.0	—	17:03:36.6			64	155	161	180	0.935	0.885
Peoria	40 42.6	-89-36.6	154	16:58:00.2	3 43.0	233	64	148	160	185	0.943	0.889
Rockford	42 16.2	-89-04.2	235	17:01:24.4			63	152	161	182	0.915	0.865
Schaumburg	42 02.0	-88-05.0	—	17:03:11.8			64	155	161	180	0.929	0.879
Skokie	42 02.0	-87-45.0	—	17:03:55.2			64	156	161	179	0.931	0.882
Springfield	39 48.0	-89-39.0	200	16:56:32.6	6 10.3	233	65	146	160	187	0.943	0.889
Urbana	40 06.3	-88-13.5	238	17:00:11.0	6 8.1	232	65	151	340	3	0.943	0.889
INDIANA												
Anderson	40 05.0	-85-50.0	—	17:05:39.5	3 13.6	231	66	159	341	357	0.943	0.890
Bloomington	39 12.6	-86-34.8	—	17:02:42.5			67	155	340	0	0.931	0.882
Evansville	37 58.8	-87-33.0	126	16:58:36.4			67	149	340	5	0.908	0.858
Fort Wayne	41 04.2	-85-09.0	259	17:08:30.7	6 9.2	231	66	163	341	354	0.943	0.890
Gary	41 35.0	-87-21.0	194	17:04:11.8	2 41.5	231	65	156	161	179	0.943	0.889
Hammond	41 37.0	-87-31.0	—	17:03:52.4	1 23.4	231	64	156	161	179	0.943	0.889
Indianapolis	39 47.4	-86-08.4	260	17:04:32.8			67	158	340	358	0.942	0.889
Muncie	40 11.5	-85-23.3	312	17:06:50.8	2 57.4	231	67	161	341	356	0.943	0.890
South Bend	41 40.0	-86-20.0	233	17:06:34.1	4 26.5	231	65	160	161	177	0.943	0.889
Terre Haute	39 28.1	-87-24.4	163	17:01:08.3	1 55.8	232	66	153	340	2	0.943	0.890
IOWA												
Ames	42 02.4	-93-36.6	—	16:51:47.4			60	139	160	191	0.881	0.827
Cedar Rapids	41 58.0	-91-39.9	240	16:55:33.9			62	144	160	187	0.900	0.849
Council Bluffs	41 16.0	-95-53.0	—	16:46:07.8			59	133	159	195	0.878	0.824
Davenport	41 32.4	-90-35.4	194	16:57:08.8			63	146	160	186	0.920	0.871
Des Moines	41 36.0	-93-37.8	308	16:51:02.6			61	138	160	191	0.892	0.839
Dubuque	42 30.0	-90-43.0	269	16:58:18.0			62	148	160	185	0.896	0.844
Iowa City	41 40.2	-91-31.8	225	16:55:23.4			62	144	160	188	0.909	0.859
Sioux City	42 30.0	-96-24.0	331	16:47:14.9			58	134	160	194	0.845	0.785
Waterloo	42 30.0	-92-22.0	279	16:54:57.9			61	143	160	188	0.882	0.828
KANSAS												
Dodge City	37 45.6	-100-01.2	847	16:32:08.9			56	118	158	206	0.916	0.866
Independence	37 13.0	-95-42.0	—	16:39:19.4	6 1.7	239	61	125	338	22	0.942	0.888
Kansas City	39 06.0	-94-39.0	246	16:44:48.4			61	131	159	197	0.941	0.888
Lawrence	38 57.6	-95-15.0	—	16:43:21.6			61	129	159	198	0.938	0.886
Overland Park	38 59.0	-94-40.0	—	16:44:34.2	1 18.1	237	61	131	159	197	0.943	0.888
Parsons	37 20.0	-95-16.0	—	16:40:24.8	5 59.5	238	61	126	338	21	0.943	0.888
Salina	38 50.1	-97-36.5	403	16:38:34.3			58	124	159	201	0.917	0.867
Topeka	39 02.4	-95-41.4	305	16:42:37.8			60	129	159	199	0.932	0.882
Wichita	37 40.8	-97-19.8	423	16:36:58.4	3 2.3	240	59	123	158	203	0.942	0.888
KENTUCKY												
Ashland	38 28.6	-82-38.4	176	17:11:26.1			69	170	341	350	0.881	0.827
Bowling Green	36 59.0	-86-27.0	167	16:59:47.2			69	151	340	4	0.873	0.818
Corbin	36 56.4	-84-06.0	—	17:05:41.7			70	161	340	356	0.852	0.793
Frankfort	38 12.0	-84-51.6	—	17:05:29.2			68	160	340	357	0.891	0.839
Lexington	38 03.6	-84-29.4	313	17:06:13.4			69	161	340	356	0.884	0.831
Louisville	38 13.2	-85-45.0	156	17:03:19.0			68	157	340	360	0.899	0.848
Owensboro	37 45.0	-87-05.0	—	16:59:23.0			68	151	340	4	0.898	0.847
Paducah	37 05.0	-88-36.0	113	16:54:42.4			67	144	339	9	0.895	0.844
LOUISIANA												
Alexandria	31 18.0	-92-28.0	—	16:35:11.4			65	118	337	30	0.791	0.719
Baton Rouge	30 27.0	-91-08.4	19	16:36:42.9			67	119	337	30	0.754	0.676
Bossier City	32 31.0	-93-42.0	—	16:34:43.6			64	118	337	29	0.834	0.772
Kenner	29 58.0	-90-15.0	—	16:37:59.5			68	120	337	30	0.732	0.650
Lafayette	30 13.2	-92-01.2	—	16:34:08.5			66	116	337	32	0.759	0.681
Lake Charles	30 12.6	-93-12.0	—	16:31:21.5			64	114	337	33	0.772	0.697
Monroe	32 30.6	-92-06.0	—	16:38:21.8			66	122	338	26	0.816	0.750
New Orleans	29 58.2	-90-04.8	2	16:38:25.6			68	120	337	29	0.730	0.647
Shreveport	32 28.2	-93-46.2	67	16:34:28.8			64	118	337	29	0.834	0.772

Table 10b
LOCAL CIRCUMSTANCES DURING THE ANNULAR SOLAR ECLIPSE OF 10 MAY 1994
FOR THE UNITED STATES OF AMERICA

Location Name	First Contact U.T. h m s	Alt	P	V	Second Contact U.T. h m s	Alt	P	V	Third Contact U.T. h m s	Alt	P	V	Fourth Contact U.T. h m s	Alt	P	V
ILLINOIS																
Des Plaines	15:24:19.7	51	249	293									18:49:28.6	62	77	52
East St. Louis	15:14:21.5	49	251	302	16:51:52.4	65	306	337	16:55:24.1	65	16	45	18:42:45.7	67	72	48
Elgin	15:23:49.1	51	248	293									18:48:37.1	63	77	53
Evanston	15:24:39.1	51	249	293									18:49:59.0	62	77	51
Joliet	15:23:14.1	51	249	294									18:48:59.4	63	76	51
Mount Prospect	15:24:18.7	51	249	293									18:49:24.0	62	77	52
Oak Lawn	15:23:55.9	51	249	293									18:49:47.4	63	76	50
Oak Park	15:24:10.9	51	249	293									18:49:41.6	62	77	51
Peoria	15:19:12.3	49	249	297	16:56:04.6	64	199	225	16:59:47.6	64	124	149	18:45:06.2	65	75	52
Rockford	15:23:05.9	50	248	292									18:46:45.7	63	78	55
Schaumburg	15:24:03.3	51	248	293									18:49:02.8	62	77	52
Skokie	15:24:33.1	51	249	293									18:49:49.6	62	77	51
Springfield	15:17:21.7	49	250	299	16:53:27.5	64	250	278	16:59:37.8	65	73	99	18:44:39.5	65	74	50
Urbana	15:20:05.5	51	251	298	16:57:07.8	65	260	284	17:03:15.9	65	64	85	18:48:16.9	64	74	47
INDIANA																
Anderson	15:23:54.5	53	252	298	17:04:07.3	66	311	328	17:07:20.9	67	13	29	18:54:02.9	62	73	41
Bloomington	15:21:04.5	53	253	301									18:51:59.6	64	72	40
Evansville	15:17:14.9	52	254	304									18:49:01.4	66	70	39
Fort Wayne	15:26:51.0	54	251	295	17:05:27.0	66	261	276	17:11:36.2	66	64	76	18:55:48.3	61	75	43
Gary	15:24:16.8	52	249	294	17:02:46.4	64	188	207	17:05:27.8	65	136	154	18:50:42.9	62	76	49
Hammond	15:24:05.4	51	249	294	17:03:05.6	64	175	194	17:04:29.0	64	149	168	18:50:19.6	62	76	50
Indianapolis	15:22:51.5	53	252	299									18:53:14.7	63	73	41
Muncie	15:24:51.3	54	252	298	17:05:26.8	66	314	330	17:08:24.2	67	11	25	18:55:07.9	62	74	41
South Bend	15:26:01.7	52	250	293	17:04:17.3	65	208	225	17:08:43.8	65	117	131	18:53:06.0	62	76	47
Terre Haute	15:20:11.2	52	252	300	17:00:15.4	66	324	346	17:02:11.2	66	360	21	18:50:03.7	64	73	43
IOWA																
Ames	15:16:43.0	46	245	293									18:35:50.7	65	78	67
Cedar Rapids	15:18:57.8	48	246	293									18:40:31.1	64	78	61
Council Bluffs	15:12:28.5	44	245	294									18:29:55.5	66	78	73
Davenport	15:19:30.2	49	248	294									18:42:59.0	64	77	57
Des Moines	15:15:45.1	46	246	294									18:35:37.3	66	78	66
Dubuque	15:21:18.8	48	246	292									18:42:54.9	64	78	60
Iowa City	15:18:31.2	48	247	294									18:40:45.2	65	77	60
Sioux City	15:14:38.1	43	243	291									18:29:17.6	65	80	77
Waterloo	15:19:12.2	47	245	292									18:38:59.3	64	79	64
KANSAS																
Dodge City	15:00:50.5	39	246	302									18:16:37.7	69	74	85
Independence	15:04:15.6	43	250	305	16:36:19.2	60	256	300	16:42:20.9	61	64	107	18:27:17.9	70	71	66
Kansas City	15:09:18.3	45	248	300									18:31:37.8	68	74	65
Lawrence	15:08:18.7	44	248	301									18:29:59.4	69	74	67
Overland Park	15:09:02.9	45	248	301	16:43:50.4	61	173	211	16:45:08.5	61	148	186	18:31:30.1	68	74	65
Parsons	15:05:00.4	44	250	305	16:37:25.9	61	259	303	16:43:25.4	62	61	103	18:28:32.9	70	71	65
Salina	15:05:27.4	42	247	300									18:23:51.1	69	75	76
Topeka	15:07:58.1	44	247	300									18:28:55.6	69	74	69
Wichita	15:03:21.5	42	248	303	16:35:23.2	59	190	235	16:38:25.5	59	129	174	18:23:29.8	70	73	74
KENTUCKY																
Ashland	15:26:57.8	57	256	302									19:01:16.8	61	70	30
Bowling Green	15:17:24.3	53	256	307									18:51:13.8	65	68	33
Corbin	15:21:41.5	55	257	307									18:57:07.9	63	68	28
Frankfort	15:22:19.2	54	255	303									18:55:49.7	63	70	33
Lexington	15:22:45.9	55	256	304									18:56:41.0	63	70	32
Louisville	15:20:45.3	53	255	304									18:53:38.7	64	70	35
Owensboro	15:17:37.6	52	254	305									18:50:04.4	65	70	37
Paducah	15:13:55.5	51	254	307									18:45:46.3	67	69	39
LOUISIANA																
Alexandria	14:58:03.1	45	259	322									18:28:27.4	76	60	41
Baton Rouge	14:58:53.8	47	261	325									18:30:46.3	76	58	31
Bossier City	14:58:12.1	44	257	318									18:26:55.3	75	63	50
Kenner	14:59:44.2	48	262	327									18:32:27.1	75	57	25
Lafayette	14:57:06.7	46	261	326									18:27:51.8	76	58	36
Lake Charles	14:55:14.9	44	260	325									18:24:30.0	77	59	43
Monroe	15:00:33.1	46	258	319									18:31:21.8	74	62	41
New Orleans	15:00:02.6	48	262	327									18:32:56.6	75	57	24
Shreveport	14:58:01.4	44	257	318									18:26:39.4	75	63	50

CIRCUMSTANCES AT MAXIMUM ECLIPSE ON 10 MAY 1994
FOR THE UNITED STATES OF AMERICA

Location Name	Latitude ° ′	Longitude ° ′	Elev. m	U.T. h m s	Umbral Durat. m s	Path Width km	Sun Alt °	Sun Az. °	P °	V °	Eclipse Mag.	Eclipse Obs.
MAINE												
Augusta	44 19.2	-69-46.2	15	17:46:04.1	6 0.1	232	60	215	166	141	0.942	0.888
Bangor	44 47.0	-68-47.0	7	17:48:10.1	5 6.0	232	59	217	167	140	0.942	0.888
Eastport	44 54.0	-67 00.0	–	17:51:52.8	5 5.2	233	57	221	167	138	0.942	0.888
Portland	43 40.2	-70-16.8	15	17:44:55.3	5 38.6	231	60	214	346	321	0.943	0.888
MARYLAND												
Annapolis	38 58.2	-76-30.0	–	17:27:54.4			68	196	343	330	0.855	0.797
Baltimore	39 18.6	-76-37.2	7	17:27:50.4			68	196	343	330	0.865	0.809
Bethesda	39 00.0	-77-10.0	–	17:26:10.6			68	194	343	332	0.859	0.802
College Park	39 00.1	-76-57.3	–	17:26:44.0			68	195	343	331	0.858	0.801
Dundalk	39 16.0	-76-31.0	–	17:28:04.7			68	196	343	330	0.863	0.807
Greenbelt	39 01.2	-76-49.6	–	17:27:05.1			68	195	343	331	0.858	0.800
Ocean City	38 23.4	-75-04.8	–	17:31:17.2			68	202	344	325	0.832	0.769
Silver Spring	39 00.0	-77 00.0	–	17:26:36.8			68	194	343	331	0.859	0.801
Wheaton	39 05.0	-77-05.0	–	17:26:27.6			68	194	343	331	0.861	0.804
MASSACHUSETTS												
Boston	42 19.2	-71-05.4	7	17:42:54.6			62	213	346	320	0.925	0.876
Brockton	42 04.0	-71-01.0	43	17:43:02.3			62	213	346	320	0.918	0.869
Brookline	42 20.0	-71-08.0	–	17:42:48.7			62	213	346	320	0.926	0.876
Cambridge	42 22.8	-71-07.8	7	17:42:49.7			62	213	346	320	0.927	0.878
Chicopee	42 10.0	-72-35.0	–	17:39:21.4			63	209	345	323	0.926	0.877
Fall River	41 42.0	-71-07.0	13	17:42:44.0			63	214	345	320	0.908	0.858
Framingham	42 16.0	-71-25.0	–	17:42:08.1			62	212	345	321	0.925	0.875
Holyoke	42 10.0	-72-40.0	38	17:39:09.6			63	208	345	323	0.926	0.877
Lawrence	42 42.0	-71-09.0	21	17:42:50.1			62	212	346	321	0.936	0.885
Lowell	42 38.0	-71-18.0	33	17:42:28.6			62	212	346	321	0.934	0.884
Lynn	42 28.0	-70-57.0	–	17:43:15.7			62	213	346	320	0.929	0.879
Malden	42 26.0	-71-04.0	–	17:42:59.1			62	213	346	320	0.928	0.879
Medford	42 25.0	-71-07.0	–	17:42:51.9			62	213	346	321	0.928	0.879
New Bedford	41 38.2	-70-55.7	5	17:43:10.3			63	214	346	319	0.906	0.855
Newton	42 21.0	-71-13.0	–	17:42:37.2			62	213	346	321	0.926	0.877
Pittsfield	42 25.0	-73-15.0	333	17:37:51.7			63	206	345	325	0.935	0.885
Quincy	42 15.0	-71 00.0	–	17:43:06.6			62	213	346	320	0.923	0.874
Somerville	42 23.0	-71-06.0	5	17:42:53.9			62	213	346	320	0.927	0.878
Springfield	42 06.6	-72-33.0	28	17:39:25.2			63	209	345	323	0.924	0.875
Waltham	42 22.0	-71-14.0	–	17:42:35.0			62	213	346	321	0.927	0.878
Weymouth	42 44.0	-70-57.0	–	17:43:18.1			62	213	346	321	0.936	0.885
Worcester	42 16.2	-71-48.6	156	17:41:12.7			63	211	345	322	0.926	0.877
MICHIGAN												
Ann Arbor	42 16.8	-83-44.4	289	17:13:09.5	5 3.1	230	65	170	162	170	0.943	0.889
Battle Creek	42 19.0	-85-11.0	269	17:09:56.5	1 29.7	230	65	165	161	173	0.943	0.889
Clinton	42 04.0	-83-58.0	–	17:12:24.3	5 34.0	230	65	169	162	170	0.943	0.889
Dearborn	42 18.0	-83-15.0	–	17:14:17.8	5 26.8	230	65	171	162	169	0.943	0.889
Dearborn Heights	41 43.0	-87-48.0	–	17:03:23.1			64	155	161	180	0.939	0.888
Detroit	42 22.8	-83-05.4	192	17:14:44.8	5 19.3	230	65	172	162	168	0.943	0.889
Farmington Hills	42 28.0	-83-23.0	–	17:14:10.2	4 39.0	230	65	171	162	169	0.943	0.889
Flint	43 01.8	-83-41.4	246	17:14:04.6			64	171	162	169	0.936	0.886
Grand Rapids	42 57.6	-85-39.6	200	17:09:38.8			64	164	161	174	0.924	0.875
Kalamazoo	42 35.0	-86 00.0	248	17:08:27.0			64	162	161	175	0.931	0.882
Lansing	42 43.2	-84-33.6	272	17:11:47.9			65	167	162	171	0.938	0.887
Livonia	42 25.0	-83-23.0	–	17:14:07.0	4 53.0	230	65	171	162	169	0.943	0.889
Mount Pleasant	43 36.0	-84-46.2	–	17:12:18.7			64	168	162	171	0.914	0.865
Pontiac	42 37.0	-83-17.0	–	17:14:33.3	3 55.9	230	65	172	162	168	0.943	0.889
Redford	42 25.0	-83-16.0	–	17:14:22.9	5 0.8	230	65	171	162	169	0.943	0.889
Roseville	42 30.0	-82-55.0	–	17:15:16.0	5 2.7	230	65	173	162	168	0.943	0.889
Royal Oak	42 29.0	-83-09.0	–	17:14:43.1	4 51.5	230	65	172	162	168	0.943	0.889
Saginaw	43 25.0	-84 00.0	195	17:13:47.9			64	170	162	170	0.924	0.875
St. Clair Shores	42 30.0	-82-54.0	–	17:15:18.3	5 3.7	230	65	173	162	168	0.943	0.889
Sault Ste. Marie	46 28.0	-84-22.0	237	17:16:04.7			61	171	163	169	0.845	0.784
Southfield	42 28.0	-83-13.0	–	17:14:32.9	4 51.3	230	65	172	162	168	0.943	0.889
Sterling Heights	42 34.0	-83-01.0	–	17:15:06.5	4 37.7	230	65	172	162	168	0.943	0.889
Taylor	42 14.0	-83-16.0	–	17:14:11.2	5 37.6	230	65	171	162	169	0.943	0.889
Troy	42 34.0	-83-09.0	–	17:14:48.3	4 26.8	230	65	172	162	168	0.943	0.889
Warren	42 33.0	-83-03.0	–	17:15:00.9	4 40.1	230	65	172	162	168	0.943	0.889

LOCAL CIRCUMSTANCES DURING THE ANNULAR SOLAR ECLIPSE OF 10 MAY 1994
FOR THE UNITED STATES OF AMERICA

Location Name	First Contact U.T. h m s	Alt °	P °	V °	Second Contact U.T. h m s	Alt °	P °	V °	Third Contact U.T. h m s	Alt °	P °	V °	Fourth Contact U.T. h m s	Alt °	P °	V °
MAINE																
Augusta	16:01:57.2	62	255	268	17:43:03.2	60	247	223	17:49:03.3	59	88	62	19:24:38.7	46	80	37
Bangor	16:04:33.5	62	255	265	17:45:34.5	59	225	199	17:50:40.5	58	111	83	19:25:43.7	45	81	38
Eastport	16:08:32.3	63	255	261	17:49:17.7	58	226	197	17:54:22.9	57	111	81	19:28:15.1	43	82	37
Portland	16:00:09.4	63	256	271	17:42:07.9	61	281	256	17:47:46.6	60	54	28	19:24:24.4	46	79	35
MARYLAND																
Annapolis	15:40:41.9	62	259	295									19:15:07.0	54	71	24
Baltimore	15:40:51.3	62	259	294									19:14:49.4	54	72	25
Bethesda	15:39:13.6	62	259	296									19:13:41.8	55	71	25
College Park	15:39:42.3	62	259	296									19:14:08.9	55	71	25
Dundalk	15:41:02.0	62	259	294									19:15:02.7	54	72	25
Greenbelt	15:40:01.1	62	259	295									19:14:25.2	55	71	25
Ocean City	15:43:20.1	64	261	295									19:18:08.9	53	70	21
Silver Spring	15:39:36.1	62	259	296									19:14:03.2	55	71	25
Wheaton	15:39:31.3	62	259	296									19:13:52.1	55	71	25
MASSACHUSETTS																
Boston	15:56:55.8	64	257	276									19:24:06.0	48	77	31
Brockton	15:56:50.1	64	258	276									19:24:24.1	48	77	30
Brookline	15:56:50.7	64	257	276									19:24:00.9	48	77	31
Cambridge	15:56:54.1	64	257	276									19:23:59.3	48	77	31
Chicopee	15:53:23.8	63	257	279									19:21:34.6	49	77	31
Fall River	15:56:13.9	64	258	278									19:24:28.1	48	76	29
Framingham	15:56:07.9	64	257	277									19:23:34.3	48	77	31
Holyoke	15:53:12.6	63	257	279									19:21:25.7	49	77	31
Lawrence	15:57:11.7	63	257	275									19:23:43.5	48	78	32
Lowell	15:56:47.3	63	257	275									19:23:31.0	48	78	32
Lynn	15:57:24.1	64	257	275									19:24:14.1	47	77	31
Malden	15:57:06.1	64	257	276									19:24:03.6	47	77	31
Medford	15:56:58.3	64	257	276									19:23:59.2	48	77	31
New Bedford	15:56:36.4	64	258	278									19:24:50.2	48	76	29
Newton	15:56:40.4	64	257	276									19:23:51.6	48	77	31
Pittsfield	15:52:12.9	62	256	280									19:20:13.4	50	77	33
Quincy	15:57:03.8	64	257	276									19:24:18.2	47	77	31
Somerville	15:56:58.4	64	257	276									19:24:02.3	48	77	31
Springfield	15:53:24.4	63	257	279									19:21:40.3	49	77	31
Waltham	15:56:39.2	64	257	276									19:23:49.2	48	77	31
Weymouth	15:57:40.8	63	257	275									19:24:02.4	47	78	32
Worcester	15:55:14.8	63	257	277									19:22:53.1	48	77	31
MICHIGAN																
Ann Arbor	15:31:26.3	55	251	291	17:10:34.9	65	218	227	17:15:38.0	65	109	116	18:58:58.6	59	77	44
Battle Creek	15:29:06.2	53	250	291	17:09:06.6	65	177	189	17:10:36.3	65	149	161	18:55:44.3	60	77	47
Clinton	15:30:40.0	54	251	291	17:09:35.0	65	227	237	17:15:09.0	65	99	107	18:58:29.5	59	76	44
Dearborn	15:32:19.0	55	251	290	17:11:31.9	65	225	233	17:16:58.7	65	102	108	19:00:03.4	59	77	43
Dearborn Heig...	15:23:51.3	51	249	293									18:49:40.4	63	76	51
Detroit	15:32:44.1	55	251	290	17:12:02.5	65	223	230	17:17:21.8	65	104	109	19:00:23.9	59	77	43
Farmington Hi...	15:32:23.2	55	251	290	17:11:47.2	65	212	220	17:16:26.2	65	115	121	18:59:44.2	59	77	44
Flint	15:32:53.4	54	250	288									18:58:57.9	59	78	46
Grand Rapids	15:29:33.8	53	249	289									18:54:36.9	60	78	49
Kalamazoo	15:28:18.3	53	249	291									18:53:52.5	61	78	49
Lansing	15:30:52.5	54	250	290									18:57:06.0	59	78	47
Livonia	15:32:17.8	55	251	290	17:11:37.3	65	215	223	17:16:30.3	65	111	117	18:59:44.7	59	77	44
Mount Pleasant	15:32:11.8	53	248	287									18:56:29.2	59	79	50
Pontiac	15:32:49.7	55	251	289	17:12:31.3	65	203	210	17:16:27.2	65	124	130	18:59:55.8	59	77	44
Redford	15:32:29.8	55	251	290	17:11:49.4	65	217	225	17:16:50.3	65	109	115	19:00:00.1	59	77	44
Roseville	15:33:15.0	55	251	289	17:12:41.6	65	218	225	17:17:44.3	65	109	114	19:00:45.2	58	77	43
Royal Oak	15:32:49.0	55	251	290	17:12:14.1	65	215	223	17:17:05.6	65	112	117	19:00:14.8	59	77	44
Saginaw	15:33:05.6	54	249	287									18:58:12.4	59	79	48
St. Clair Sho...	15:33:16.8	55	251	289	17:12:43.4	65	218	225	17:17:47.1	65	109	113	19:00:47.4	58	77	43
Sault Ste. Ma...	15:38:22.9	53	245	279									18:56:20.3	57	84	57
Southfield	15:32:40.3	55	251	290	17:12:04.0	65	215	223	17:16:55.3	65	112	117	19:00:06.2	59	77	44
Sterling Heig...	15:33:11.8	55	251	289	17:12:44.1	65	212	219	17:17:21.8	65	115	120	19:00:31.4	58	77	44
Taylor	15:32:10.1	55	251	290	17:11:20.3	65	228	237	17:16:57.8	65	98	104	19:00:01.7	59	77	43
Troy	15:32:58.0	55	251	289	17:12:31.2	65	209	217	17:16:58.1	65	118	123	19:00:13.9	58	77	44
Warren	15:33:06.5	55	251	289	17:12:37.4	65	212	219	17:17:17.5	65	115	120	19:00:27.2	58	77	44

CIRCUMSTANCES AT MAXIMUM ECLIPSE ON 10 MAY 1994
FOR THE UNITED STATES OF AMERICA

Location Name	Latitude ° ′	Longitude ° ′	Elev. m	U.T. h m s	Umbral Durat. m s	Path Width km	Sun Alt °	Sun Az. °	P °	V °	Eclipse Mag.	Eclipse Obs.
MICHIGAN												
Westland	42 19.0	-83-24.0	—	17:13:58.3	5 15.7	230	65	171	162	169	0.943	0.889
Wyoming	42 54.0	-85-42.0	—	17:09:29.2			64	164	161	174	0.925	0.876
MINNESOTA												
Bloomington	44 50.0	-93-18.0	—	16:56:40.1			59	145	161	186	0.819	0.753
Duluth	46 47.4	-92-06.6	200	17:01:36.5			58	151	161	182	0.783	0.710
Hibbing	47 25.2	-92-55.2	—	17:01:00.5			57	149	161.	183	0.763	0.686
Internat'l Falls	48 36.0	-93-24.6	—	17:01:44.3			56	150	162	182	0.732	0.650
Mankato	44 09.6	-94 00.0	—	16:54:21.2			59	142	160	188	0.829	0.765
Minneapolis	44 57.6	-93-16.2	274	16:56:54.6			59	145	161	186	0.816	0.750
Northfield	44 27.6	-93-09.6	—	16:56:22.8			59	145	160	186	0.829	0.765
Rochester	44 01.0	-92-30.0	—	16:56:58.8			60	146	160	186	0.845	0.784
St. Cloud	45 34.0	-94-10.4	341	16:56:08.9			58	144	161	187	0.795	0.724
St. Paul	44 57.0	-93-05.0	256	16:57:14.6			59	146	161	186	0.818	0.752
MISSISSIPPI												
Aberdeen	33 49.0	-88-33.0	—	16:49:22.2			69	136	338	16	0.812	0.745
Biloxi	30 24.6	-88-55.2	7	16:42:14.9			70	125	337	26	0.729	0.645
Greenville	33 25.0	-91 00.0	—	16:42:39.0			67	127	338	22	0.827	0.763
Jackson	32 19.2	-90-12.0	98	16:42:33.9			68	127	338	24	0.791	0.720
Meridian	32 21.0	-88-41.0	—	16:46:26.1			69	131	338	20	0.776	0.702
Vicksburg	32 20.0	-90-50.0	—	16:41:02.6			67	125	338	25	0.798	0.729
MISSOURI												
Cape Girardeau	37 18.6	-89-31.8	—	16:52:53.0			66	141	339	11	0.909	0.860
Columbia	38 55.0	-92-19.0	240	16:49:19.8	6 4.5	235	63	137	159	194	0.943	0.889
Fayette	39 09.0	-92-42.0	—	16:48:54.6	5 26.8	235	63	136	159	194	0.943	0.889
Florissant	38 47.0	-90-20.0	—	16:53:26.1	4 46.0	234	65	142	339	10	0.943	0.889
Independence	39 06.0	-94-26.0	—	16:45:14.7	0 43.9	237	61	132	159	197	0.943	0.888
Jefferson City	38 34.2	-92-10.8	—	16:49:02.6	6 0.3	235	63	136	339	14	0.943	0.889
Kansas City	39 05.0	-94-35.0	243	16:44:54.7			61	131	159	197	0.942	0.888
Mexico	39 10.0	-91-53.0	—	16:50:40.3	6 0.7	235	63	138	159	192	0.943	0.889
Nevada	37 51.0	-94-22.0	—	16:43:11.2	6 4.8	237	62	129	339	19	0.943	0.889
St. Joseph	39 44.0	-94-49.0	279	16:45:34.2			61	132	159	196	0.924	0.875
St. Louis	38 37.8	-90-15.0	149	16:53:22.5	3 44.4	234	65	142	339	10	0.943	0.889
Sedalia	38 42.0	-93-14.0	—	16:47:01.7	5 54.1	236	62	134	159	196	0.943	0.889
Springfield	37 12.0	-93-17.4	427	16:44:17.7	1 15.3	237	63	130	339	18	0.943	0.889
MONTANA												
Billings	45 46.8	-108-32.4	1024	16:34:12.1			47	118	160	201	0.664	0.569
Bozeman	45 41.0	-111 00.0	—	16:30:57.2			45	114	160	202	0.644	0.545
Butte	46 00.0	-112-31.0	1891	16:29:45.6			44	113	160	203	0.624	0.522
Great Falls	47 30.0	-111-15.0	1096	16:34:03.5			45	117	160	200	0.608	0.503
Helena	46 35.4	-112-01.8	1363	16:31:26.5			44	114	160	202	0.617	0.514
Missoula	46 51.6	-114 00.0	1047	16:29:44.7			43	112	160	202	0.595	0.489
NEBRASKA												
Grand Island	40 55.8	-98-21.0	—	16:40:59.5			57	127	159	199	0.862	0.805
Lincoln	40 48.6	-96-40.2	377	16:43:51.6			59	130	159	197	0.881	0.827
North Platte	41 08.0	-100-45.0	—	16:37:11.0			55	123	159	201	0.834	0.771
Omaha	41 18.0	-95-57.0	341	16:46:03.5			59	132	159	195	0.877	0.822
Scottsbluff	41 51.6	-103-39.6	—	16:33:49.7			52	119	159	202	0.790	0.718
NEVADA												
Carson City	39 09.0	-119-46.8	1535	16:08:58.1			37	97	159	214	0.680	0.587
Las Vegas	36 10.2	-115-10.2	709	16:07:37.6			40	97	159	216	0.784	0.711
Reno	39 31.5	-119-48.7	1445	16:09:40.3			37	97	160	213	0.673	0.578
NEW HAMPSHIRE												
Concord	43 10.0	-71-30.0	95	17:42:06.7	4 2.7	231	62	211	346	322	0.943	0.889
Hanover	43 42.3	-72-17.0	—	17:40:26.6	6 5.1	231	62	208	345	324	0.943	0.889
Manchester	42 59.4	-71-27.6	57	17:42:10.3	2 16.4	231	62	211	346	322	0.943	0.889
Nashua	42 47.0	-71-23.0	—	17:42:18.6			62	212	346	321	0.939	0.887
NEW JERSEY												
Atlantic City	39 21.6	-74-26.4	3	17:33:33.5			67	204	344	324	0.856	0.797
Bayonne	40 40.0	-74-07.0	—	17:35:03.0			65	205	344	325	0.890	0.838
Camden	39 56.0	-75-06.0	10	17:32:10.6			66	202	344	326	0.875	0.820
Cape May	38 56.4	-74-54.6	—	17:32:05.0			67	203	344	325	0.846	0.786
Cherry Hill	39 56.0	-75-01.0	—	17:32:23.4			66	202	344	326	0.874	0.819
Clifton	40 35.0	-74-09.0	—	17:34:55.6			65	205	344	325	0.888	0.836
East Orange	40 46.0	-74-12.0	—	17:34:53.4			65	204	344	325	0.894	0.842
Edison	40 27.0	-74-18.0	—	17:34:29.2			66	204	344	325	0.885	0.832
Elizabeth	40 40.0	-74-13.0	7	17:34:48.0			65	204	344	325	0.891	0.839
Irvington	40 43.0	-74-15.0	—	17:34:44.4			65	204	344	325	0.892	0.841
Jersey City	40 43.0	-74-05.0	7	17:35:09.4			65	205	344	325	0.892	0.840

LOCAL CIRCUMSTANCES DURING THE ANNULAR SOLAR ECLIPSE OF 10 MAY 1994
FOR THE UNITED STATES OF AMERICA

Location Name	First Contact				Second Contact				Third Contact				Fourth Contact			
	U.T. h m s	Alt °	P °	V °	U.T. h m s	Alt °	P °	V °	U.T. h m s	Alt °	P °	V °	U.T. h m s	Alt °	P °	V °
MICHIGAN																
Westland	15:32:05.3	55	251	290	17:11:17.7	65	221	230	17:16:33.4	65	105	111	18:59:43.4	59	77	43
Wyoming	15:29:23.1	53	249	290									18:54:31.8	60	78	49
MINNESOTA																
Bloomington	15:23:11.4	46	242	285									18:37:14.5	62	83	72
Duluth	15:28:55.4	47	241	280									18:39:52.3	60	85	73
Hibbing	15:29:27.6	46	239	278									18:38:02.1	60	87	76
Internat'l Fa..	15:31:40.3	46	238	275									18:36:47.5	59	89	80
Mankato	15:20:54.3	45	243	287									18:35:31.6	63	82	73
Minneapolis	15:23:30.4	46	242	285									18:37:19.4	62	83	72
Northfield	15:22:31.2	46	243	286									18:37:31.2	63	82	71
Rochester	15:22:18.9	47	244	287									18:38:59.1	63	81	68
St. Cloud	15:23:52.9	45	241	283									18:35:19.4	62	84	76
St. Paul	15:23:41.7	46	242	285									18:37:44.8	62	83	72
MISSISSIPPI																
Aberdeen	15:08:29.8	50	259	317									18:42:52.4	70	63	29
Biloxi	15:02:45.8	50	263	327									18:37:00.9	74	57	20
Greenville	15:03:45.5	48	257	317									18:35:39.8	72	63	38
Jackson	15:03:18.1	48	259	320									18:36:23.2	73	61	31
Meridian	15:05:59.4	50	260	321									18:40:39.1	72	61	25
Vicksburg	15:02:16.4	48	259	320									18:34:38.3	73	61	34
MISSOURI																
Cape Girardeau	15:12:51.2	50	253	306									18:43:31.5	68	69	43
Columbia	15:11:53.3	47	250	301	16:46:17.1	63	243	279	16:52:21.6	63	78	112	18:37:27.8	68	73	56
Fayette	15:11:50.9	46	249	301	16:46:09.0	62	224	259	16:51:35.9	63	98	131	18:36:38.2	68	73	58
Florissant	15:14:24.0	49	251	302	16:51:06.4	64	291	322	16:55:52.4	65	31	61	18:42:25.6	67	72	49
Independence	15:09:34.1	45	248	300	16:44:47.9	61	167	205	16:45:31.8	61	154	191	18:32:11.0	68	74	64
Jefferson City	15:11:23.3	47	250	302	16:46:03.6	63	263	298	16:52:03.9	64	59	92	18:37:34.7	68	72	55
Kansas City	15:09:21.1	45	248	300									18:31:47.3	68	74	65
Mexico	15:12:58.1	47	250	301	16:47:39.1	63	239	274	16:53:39.8	64	82	114	18:38:43.3	67	73	55
Nevada	15:07:07.8	45	250	304	16:40:09.2	61	253	294	16:46:14.0	62	67	106	18:31:22.0	69	72	62
St. Joseph	15:10:25.0	44	247	299									18:31:39.0	68	75	67
St. Louis	15:14:13.6	49	251	302	16:51:34.4	65	304	335	16:55:18.8	65	18	48	18:42:32.9	67	72	49
Sedalia	15:10:15.4	46	249	302	16:44:03.5	62	236	273	16:49:57.6	63	85	121	18:34:58.1	68	73	59
Springfield	15:07:13.9	46	251	306	16:43:44.7	63	328	8	16:44:60.0	63	352	31	18:33:37.0	70	70	56
MONTANA																
Billings	15:13:33.0	34	233	279									18:03:12.5	59	90	113
Bozeman	15:12:17.6	32	232	279									17:57:54.2	57	91	118
Butte	15:12:33.7	31	231	277									17:54:58.5	56	92	121
Great Falls	15:16:46.6	32	230	274									17:58:40.2	56	94	119
Helena	15:14:12.1	31	230	276									17:56:25.9	56	93	120
Missoula	15:14:18.3	30	229	275									17:52:39.1	54	94	124
NEBRASKA																
Grand Island	15:09:13.4	41	244	295									18:23:37.2	67	78	81
Lincoln	15:10:38.8	43	245	295									18:27:42.4	67	77	75
North Platte	15:07:31.2	39	242	293									18:17:51.8	66	79	90
Omaha	15:12:28.4	44	245	294									18:29:46.8	66	78	73
Scottsbluff	15:06:57.0	37	240	290									18:11:24.6	64	82	99
NEVADA																
Carson City	14:54:49.9	22	234	288									17:33:09.4	52	87	135
Las Vegas	14:49:01.1	25	239	297									17:38:30.8	58	80	129
Reno	14:55:40.5	23	233	287									17:33:34.0	52	88	135
NEW HAMPSHIRE																
Concord	15:56:55.7	63	256	274	17:40:09.3	62	306	283	17:44:12.0	61	28	4	19:22:46.8	48	78	34
Hanover	15:55:51.2	62	255	274	17:37:24.8	62	266	246	17:43:29.9	61	68	46	19:21:01.3	48	79	36
Manchester	15:56:49.2	63	256	275	17:41:07.0	62	326	302	17:43:23.4	62	9	345	19:22:58.9	48	78	33
Nashua	15:56:45.8	63	256	275									19:23:15.9	48	78	33
NEW JERSEY																
Atlantic City	15:45:56.5	64	260	291									19:19:17.1	52	72	24
Bayonne	15:48:11.7	63	258	286									19:19:29.8	51	74	28
Camden	15:45:04.6	63	259	290									19:17:48.8	52	73	26
Cape May	15:44:21.5	64	260	293									19:18:25.4	53	71	23
Cherry Hill	15:45:16.1	63	259	290									19:17:58.8	52	73	26
Clifton	15:48:01.3	63	258	287									19:19:27.9	51	74	27
East Orange	15:48:07.4	63	258	286									19:19:17.7	51	74	28
Edison	15:47:31.3	63	258	287									19:19:13.5	51	74	27
Elizabeth	15:47:58.0	63	258	286									19:19:18.2	51	74	28
Irvington	15:47:57.0	63	258	286									19:19:13.1	51	74	28
Jersey City	15:48:19.8	63	258	286									19:19:32.4	51	74	28

Table 10a
CIRCUMSTANCES AT MAXIMUM ECLIPSE ON 10 MAY 1994
FOR THE UNITED STATES OF AMERICA

Location Name	Latitude ° ′	Longitude ° ′	Elev. m	U.T. h m s	Umbral Durat. m s	Path Width km	Sun Alt °	Sun Az. °	P °	V °	Eclipse Mag.	Eclipse Obs.
NEW JERSEY												
Newark	40 44.4	-74-11.4	–	17:34:54.1			65	204	344	325	0.893	0.841
Passaic	40 52.0	-74-08.0	–	17:35:06.1			65	204	344	325	0.896	0.845
Paterson	40 55.0	-74-10.0	33	17:35:02.6			65	204	344	325	0.898	0.846
Princeton	40 21.0	-74-39.6	–	17:33:31.6			66	203	344	326	0.884	0.831
Trenton	40 13.2	-74-45.6	11	17:33:12.2			66	203	344	326	0.881	0.827
Union	40 41.0	-74-15.0	–	17:34:43.5			65	204	344	325	0.891	0.839
Union City	40 46.0	-74-01.0	–	17:35:20.8			65	205	344	325	0.893	0.841
Vineland	39 30.0	-75 00.0	–	17:32:10.9			67	202	344	326	0.862	0.805
NEW MEXICO												
Alamagordo	32 54.0	-105-57.0	–	16:12:50.9	5 0.3	254	50	102	157	217	0.941	0.885
Albuquerque	35 05.0	-106-40.0	1742	16:16:07.3			49	104	158	214	0.900	0.848
Clovis	34 24.0	-103-12.0	–	16:20:08.8	4 56.6	249	53	107	158	214	0.941	0.886
Deming	32 16.0	-107-45.0	–	16:08:59.4	3 52.8	256	47	99	157	219	0.940	0.884
Las Cruces	32 20.4	-106-43.8	–	16:10:35.1	5 16.1	255	49	100	157	218	0.941	0.885
Portales	34 11.0	-103-20.0	–	16:19:29.9	5 17.3	249	53	107	158	214	0.941	0.886
Roswell	33 23.0	-104-32.0	–	16:15:58.9	5 32.1	252	51	104	157	216	0.941	0.886
Santa Fe	35 40.2	-105-57.0	2280	16:18:20.4			50	106	158	213	0.896	0.843
Sunspot	32 47.2	-105-49.2	–	16:12:49.0	5 22.0	254	50	102	157	217	0.941	0.885
NEW YORK												
Albany	42 39.6	-73-46.8	7	17:36:42.0	1 32.4	230	63	205	345	326	0.943	0.889
Binghamton	42 05.0	-75-55.0	284	17:31:22.4			65	198	344	330	0.937	0.887
Buffalo	42 54.6	-78-51.0	231	17:24:59.7	6 11.1	229	65	187	163	157	0.943	0.889
Cheektowaga	42 54.0	-78-46.0	–	17:25:10.8	6 12.0	229	65	188	163	157	0.943	0.889
Irondequoit	43 12.0	-77-36.0	–	17:28:04.8	6 8.7	229	64	192	164	155	0.943	0.889
Ithaca	42 26.4	-76-29.4	–	17:30:12.6	3 58.8	230	65	196	344	332	0.943	0.889
Jamestown	42 06.6	-79-14.4	–	17:23:28.0	5 6.2	229	65	186	343	338	0.943	0.890
Mount Vernon	40 55.0	-73-51.0	–	17:35:49.7			65	205	344	324	0.896	0.845
New Rochelle	40 55.0	-73-47.0	–	17:35:59.6			65	206	344	324	0.896	0.845
New York	40 43.8	-73-55.2	43	17:35:34.3			65	205	344	324	0.891	0.839
Niagara Falls	43 06.0	-79-02.0	187	17:24:42.7	5 54.2	229	64	187	163	158	0.943	0.889
Poughkeepsie	41 42.0	-73-55.2	–	17:35:59.5			64	205	345	325	0.918	0.869
Rochester	43 09.6	-77-36.6	169	17:28:02.0	6 10.4	229	64	192	164	155	0.943	0.889
Schenectady	42 47.0	-73-53.0	80	17:36:30.3	3 24.5	230	63	204	345	326	0.943	0.889
Syracuse	43 05.0	-76-10.0	131	17:31:19.6	6 4.0	230	64	197	344	331	0.943	0.889
Tonawanda	43 01.0	-78-53.0	–	17:24:59.7	6 4.9	229	65	187	163	158	0.943	0.889
Troy	42 45.0	-73-45.0	11	17:36:48.2	2 52.6	230	63	205	345	326	0.943	0.889
Utica	43 06.2	-75-13.6	136	17:33:30.9	5 50.3	230	63	200	344	329	0.943	0.889
West Seneca	42 50.0	-78-45.0	–	17:25:10.2	6 13.1	229	65	188	163	157	0.943	0.889
Yonkers	40 57.0	-73-54.0	3	17:35:43.1			65	205	344	324	0.897	0.846
NORTH CAROLINA												
Asheville	35 35.4	-82-33.6	702	17:07:57.7			72	166	340	353	0.804	0.735
Charlotte	35 13.2	-80-49.8	236	17:12:17.8			72	174	341	346	0.780	0.707
Durham	36 00.0	-78-54.6	133	17:18:38.4			72	185	342	337	0.788	0.716
Fayetteville	35 02.0	-78-54.0	–	17:17:36.8			73	184	341	338	0.761	0.684
Greensboro	36 04.2	-79-48.6	275	17:16:11.0			72	180	341	341	0.796	0.726
High Point	35 55.0	-80 00.0	–	17:15:28.4			72	179	341	342	0.793	0.722
Raleigh	35 47.4	-78-39.0	120	17:19:09.4			72	186	342	336	0.780	0.707
Wilmington	34 13.2	-77-55.8	9	17:19:34.9			73	190	342	333	0.732	0.649
Winston-Salem	36 06.0	-80-15.6	282	17:14:57.6			72	178	341	343	0.800	0.731
NORTH DAKOTA												
Bismarck	46 48.6	-100-46.8	540	16:47:03.0			53	132	160	193	0.712	0.625
Fargo	46 52.2	-96-47.4	295	16:53:33.8			55	140	161	188	0.745	0.664
Grand Forks	47 55.0	-97-05.0	–	16:54:39.8			55	141	161	187	0.719	0.634
Minot	48 14.4	-101-18.0	509	16:48:36.8			52	133	161	192	0.678	0.585
OHIO												
Akron	41 05.0	-81-30.7	287	17:17:06.9			67	177	342	345	0.942	0.889
Canton	40 50.0	-81-25.0	338	17:17:05.5			67	177	342	344	0.935	0.885
Cincinnati	39 08.4	-84-30.6	180	17:07:35.7			68	163	341	355	0.913	0.863
Cleveland	41 28.8	-81-39.6	217	17:17:09.4	4 47.2	230	66	177	342	345	0.943	0.890
Cleveland Heights	41 30.0	-81-35.0	–	17:17:21.5	4 47.3	230	66	177	342	345	0.943	0.890
Columbus	39 58.8	-82-59.4	256	17:12:20.3			67	170	341	349	0.923	0.874
Dayton	39 45.0	-84-15.0	188	17:09:00.1			67	165	341	353	0.926	0.878
Elyria	41 22.0	-82-07.0	–	17:15:57.5	4 47.4	230	66	175	342	346	0.943	0.890
Euclid	41 34.0	-81-32.0	–	17:17:32.6	5 2.1	230	66	177	342	344	0.943	0.890
Hamilton	39 22.0	-84-33.0	197	17:07:47.4			68	163	341	355	0.919	0.870

58

LOCAL CIRCUMSTANCES DURING THE ANNULAR SOLAR ECLIPSE OF 10 MAY 1994 FOR THE UNITED STATES OF AMERICA

Location Name	First Contact U.T. h m s	Alt °	P °	V °	Second Contact U.T. h m s	Alt °	P °	V °	Third Contact U.T. h m s	Alt °	P °	V °	Fourth Contact U.T. h m s	Alt °	P °	V °
NEW JERSEY																
Newark	15:48:06.9	63	258	286									19:19:19.5	51	74	28
Passaic	15:48:23.6	63	258	286									19:19:22.9	51	74	28
Paterson	15:48:22.6	63	258	286									19:19:17.8	51	74	29
Princeton	15:46:34.9	63	258	288									19:18:33.4	52	74	27
Trenton	15:46:11.9	63	258	289									19:18:24.2	52	73	27
Union	15:47:54.7	63	258	286									19:19:13.9	51	74	28
Union City	15:48:32.4	63	258	286									19:19:38.8	51	74	28
Vineland	15:44:47.6	63	259	291									19:18:07.5	52	72	25
NEW MEXICO																
Alamagordo	14:46:25.9	32	248	310	16:10:19.1	49	219	280	16:15:19.4	50	98	157	17:54:53.8	69	70	111
Albuquerque	14:50:14.9	32	245	305									17:56:27.2	67	73	111
Clovis	14:51:19.1	35	248	308	16:17:38.5	52	217	273	16:22:35.1	53	101	157	18:04:10.8	71	70	100
Deming	14:44:07.8	30	248	310	16:07:00.5	47	202	263	16:10:53.3	48	115	177	17:49:25.7	67	70	117
Las Cruces	14:44:52.5	31	248	311	16:07:55.8	48	226	287	16:13:11.9	49	91	152	17:52:02.6	68	69	114
Portales	14:50:47.5	35	248	309	16:16:49.7	52	224	281	16:22:07.1	53	93	150	18:03:31.1	71	70	100
Roswell	14:48:20.2	33	249	310	16:13:11.9	51	232	291	16:18:44.0	52	85	144	17:59:13.9	70	70	106
Santa Fe	14:51:52.9	33	245	304									17:59:02.7	67	74	108
Sunspot	14:46:18.2	32	248	310	16:10:06.8	49	228	288	16:15:28.9	50	89	149	17:55:02.2	69	69	111
NEW YORK																
Albany	15:51:21.8	62	256	280	17:36:00.9	63	332	313	17:37:33.4	63	1	342	19:19:06.1	50	77	34
Binghamton	15:46:03.3	61	255	284									19:15:23.6	52	76	34
Buffalo	15:41:22.8	58	253	285	17:21:53.7	65	249	245	17:28:04.8	65	80	73	19:09:11.9	55	78	39
Cheektowaga	15:41:31.5	58	253	285	17:22:04.4	65	251	246	17:28:16.4	65	79	72	19:09:22.1	55	78	39
Irondequoit	15:44:15.2	59	253	283	17:24:59.8	64	247	240	17:31:08.5	64	83	73	19:11:33.1	53	78	39
Ithaca	15:45:21.7	60	254	284	17:28:17.4	65	306	294	17:32:16.2	64	25	12	19:14:06.1	53	77	35
Jamestown	15:39:21.9	58	253	288	17:20:58.0	66	290	286	17:26:04.2	65	39	34	19:08:41.7	55	76	37
Mount Vernon	15:49:05.7	63	258	285									19:19:54.1	51	75	28
New Rochelle	15:49:14.8	63	258	285									19:20:01.7	51	75	28
New York	15:48:43.0	63	258	286									19:19:50.9	51	74	28
Niagara Falls	15:41:20.0	58	252	285	17:21:44.0	65	237	233	17:27:38.2	64	93	86	19:08:44.8	55	78	40
Poughkeepsie	15:49:52.3	63	257	283									19:19:23.5	51	76	31
Rochester	15:44:10.4	59	253	283	17:24:56.3	64	249	242	17:31:06.7	64	81	71	19:11:33.3	53	78	38
Schenectady	15:51:17.8	62	255	279	17:34:52.4	63	313	295	17:38:17.0	63	20	0	19:18:50.2	50	78	34
Syracuse	15:46:56.3	60	254	282	17:28:18.7	64	268	256	17:34:22.7	64	63	49	19:14:23.5	52	78	37
Tonawanda	15:41:29.2	58	252	285	17:21:56.2	65	243	239	17:28:01.1	64	86	79	19:09:05.0	55	78	39
Troy	15:51:32.4	62	255	279	17:35:26.6	63	319	300	17:38:19.2	63	14	355	19:19:06.0	50	78	34
Utica	15:48:53.5	61	254	280	17:30:37.5	64	276	262	17:36:27.8	63	56	40	19:16:09.4	51	78	36
West Seneca	15:41:27.1	58	253	285	17:22:03.6	65	255	250	17:28:16.7	65	75	68	19:09:25.8	55	77	39
Yonkers	15:49:01.2	63	258	285									19:19:47.5	51	75	28
NORTH CAROLINA																
Asheville	15:22:46.2	57	260	311									19:00:13.0	63	65	20
Charlotte	15:26:02.8	59	262	312									19:04:17.6	61	65	16
Durham	15:31:26.6	61	262	308									19:09:19.0	58	66	17
Fayetteville	15:30:19.0	62	263	311									19:08:53.5	59	64	13
Greensboro	15:29:27.4	60	261	308									19:07:12.6	59	66	18
High Point	15:28:49.7	60	261	309									19:06:40.9	60	66	18
Raleigh	15:31:48.1	62	263	308									19:09:50.7	58	66	16
Wilmington	15:31:49.0	63	265	313									19:10:45.9	58	63	10
Winston-Salem	15:28:28.8	60	261	308									19:06:08.6	60	66	19
NORTH DAKOTA																
Bismarck	15:20:48.0	40	236	279									18:20:44.9	61	88	95
Fargo	15:24:17.9	43	238	280									18:29:35.5	61	87	85
Grand Forks	15:26:31.6	43	237	277									18:29:01.2	60	89	87
Minot	15:23:55.8	40	234	276									18:19:58.6	59	90	98
OHIO																
Akron	15:33:21.4	57	254	293									19:04:03.1	58	75	37
Canton	15:33:07.5	57	254	294									19:04:17.1	58	74	36
Cincinnati	15:24:32.9	55	254	300									18:56:59.4	62	72	36
Cleveland	15:33:44.4	57	253	292	17:14:49.3	66	294	298	17:19:36.5	66	34	35	19:03:40.5	58	75	38
Cleveland Hei...	15:33:55.0	57	253	292	17:15:01.4	66	294	297	17:19:48.6	66	34	35	19:03:50.3	58	75	38
Columbus	15:28:45.2	56	254	297									19:00:43.1	60	73	36
Dayton	15:26:03.9	55	254	299									18:57:44.6	61	73	37
Elyria	15:32:42.5	56	253	293	17:13:37.3	66	294	299	17:18:24.7	66	33	37	19:02:40.8	58	75	39
Euclid	15:34:07.3	57	253	292	17:15:04.8	66	290	294	17:20:06.8	66	37	38	19:03:56.2	58	75	38
Hamilton	15:24:52.0	55	254	300									18:56:57.1	62	72	36

CIRCUMSTANCES AT MAXIMUM ECLIPSE ON 10 MAY 1994
FOR THE UNITED STATES OF AMERICA

Location Name	Latitude ° '	Longitude ° '	Elev. m	U.T. h m s	Umbral Durat. m s	Path Width km	Sun Alt °	Sun Az. °	P °	V °	Eclipse Mag.	Eclipse Obs.
OHIO												
Kettering	39 40.0	-84-15.0	—	17:08:53.8			67	165	341	353	0.924	0.875
Lakewood	41 29.0	-81-48.0	—	17:16:49.7	4 57.6	230	66	176	342	345	0.943	0.890
Lima	40 45.0	-84-06.0	284	17:10:34.6	4 21.1	230	66	167	341	352	0.943	0.890
Lorain	41 28.0	-82-10.0	200	17:15:56.6	5 16.1	230	66	175	342	346	0.943	0.890
Mansfield	40 45.0	-82-30.0	—	17:14:23.3			67	173	342	347	0.940	0.888
Parma	41 23.0	-81-44.0	—	17:16:53.1	4 21.6	230	66	176	342	345	0.943	0.890
Springfield	39 55.0	-83-50.0	322	17:10:12.9			67	167	341	352	0.928	0.879
Steubenville	40 22.0	-80-37.0	217	17:18:35.1			67	180	342	342	0.917	0.868
Toledo	41 40.2	-83-34.2	192	17:12:52.5	6 13.5	230	66	170	342	350	0.943	0.890
Warren	41 15.0	-80-50.0	—	17:18:54.5			66	180	342	343	0.942	0.889
Youngstown	41 05.4	-80-39.0	276	17:19:12.0			67	180	342	342	0.937	0.886
OKLAHOMA												
Clinton	35 31.0	-98-59.0	—	16:29:44.3	5 47.9	243	58	115	338	29	0.942	0.887
Enid	36 23.7	-97-52.5	407	16:33:30.9	5 59.6	241	59	119	158	206	0.942	0.888
Lawton	34 36.0	-98-25.0	—	16:29:01.6			58	114	338	30	0.937	0.885
Midwest City	35 26.0	-97-23.0	—	16:32:38.1	2 23.4	242	59	118	338	27	0.942	0.888
Norman	35 13.0	-97-25.0	—	16:32:09.2			59	117	338	28	0.940	0.887
Muskogee	35 44.0	-95-21.0	—	16:37:17.5			61	122	338	24	0.930	0.880
Oklahoma City	35 28.8	-97-31.8	422	16:32:26.1	3 25.6	242	59	118	338	27	0.942	0.888
Ponca City	36 42.0	-97-05.0	—	16:35:37.4	6 1.2	240	59	121	338	25	0.942	0.888
Tulsa	36 08.4	-95-56.4	264	16:36:50.6	2 45.8	240	61	122	338	24	0.942	0.888
OREGON												
Burns	43 35.0	-119-05.0	—	16:18:16.9			38	102	160	208	0.607	0.502
Corvallis	44 34.0	-123-16.0	—	16:16:39.0			35	100	161	208	0.552	0.441
Eugene	44 03.0	-123-06.0	138	16:15:46.2			35	99	160	209	0.562	0.452
Medford	42 19.0	-122-52.0	—	16:12:34.5			35	98	160	211	0.594	0.488
Pendleton	45 40.2	-118-48.0	—	16:22:36.9			39	105	160	206	0.573	0.464
Portland	45 31.2	-122-39.0	7	16:18:59.8			35	101	161	207	0.542	0.429
Salem	44 55.8	-123-01.8	51	16:17:32.6			35	100	161	208	0.548	0.436
PENNSYLVANIA												
Allentown	40 35.0	-75-30.0	84	17:31:32.4			66	200	344	328	0.894	0.843
Altoona	40 25.0	-78-25.0	387	17:24:06.4			67	189	343	336	0.905	0.855
Bethlehem	40 40.0	-75-25.0	77	17:31:47.8			66	200	344	328	0.896	0.845
Erie	42 07.2	-80-04.8	225	17:21:29.4	5 44.1	229	66	183	343	340	0.943	0.890
Harrisburg	40 16.2	-76-52.8	120	17:27:51.7			67	195	343	331	0.893	0.841
Lancaster	40 05.0	-76-20.0	116	17:29:07.2			67	197	343	330	0.885	0.832
Penn Hills	40 28.0	-79-51.0	—	17:20:34.6			67	183	342	340	0.915	0.866
Philadelphia	40 00.0	-75-09.0	33	17:32:05.2			66	201	344	327	0.877	0.822
Pittsburgh	40 26.4	-79-58.2	245	17:20:15.3			67	182	342	340	0.915	0.866
Reading	40 20.0	-75-55.0	87	17:30:20.4			66	198	344	329	0.890	0.837
Scranton	41 24.6	-75-40.2	238	17:31:35.6			65	199	344	329	0.918	0.869
Upper Darby	39 58.0	-75-16.0	—	17:31:46.2			67	201	344	327	0.876	0.822
Wilkes-Barre	41 14.5	-75-53.3	210	17:30:57.7			66	198	344	329	0.914	0.865
RHODE ISLAND												
Cranston	41 46.0	-71-25.0	—	17:42:01.9			63	213	345	320	0.911	0.861
East Providence	41 49.0	-71-22.0	—	17:42:09.7			63	213	345	320	0.912	0.862
Pawtucket	41 53.0	-71-23.0	—	17:42:08.1			63	213	345	320	0.914	0.865
Providence	41 49.2	-71-25.8	—	17:42:00.6			63	212	345	320	0.912	0.863
Warwick	41 42.0	-71-27.0	26	17:41:56.2			63	213	345	320	0.909	0.859
SOUTH CAROLINA												
Charleston	32 48.6	-79-57.6	3	17:11:45.5			75	176	341	344	0.707	0.620
Columbia	34 00.6	-81 00.0	62	17:10:15.1			74	171	340	348	0.748	0.669
Greenville	34 51.0	-82-23.4	317	17:07:25.4			72	165	340	353	0.782	0.710
North Charleston	32 49.0	-79-57.0	—	17:11:47.9			75	176	341	344	0.707	0.620
Spartanburg	34 56.4	-81-55.8	287	17:08:50.0			72	168	340	351	0.781	0.708
SOUTH DAKOTA												
Pierre	44 22.2	-100-20.4	486	16:43:37.1			54	129	160	196	0.768	0.692
Rapid City	44 04.2	-103-13.8	1060	16:38:32.1			52	123	160	199	0.747	0.667
Sioux Falls	43 32.4	-96-42.6	364	16:48:24.8			57	135	160	193	0.819	0.753
TENNESSEE												
Chattanooga	35 02.4	-85-16.8	221	16:59:50.2			71	152	339	4	0.812	0.745
Clarksville	36 30.0	-87-23.0	—	16:56:43.8			68	147	339	7	0.869	0.813
Knoxville	35 58.8	-83-56.4	292	17:04:46.6			71	160	340	357	0.825	0.761
Memphis	35 07.2	-89-59.4	90	16:48:06.4			67	135	338	17	0.859	0.802
Nashville	36 09.6	-86-46.2	194	16:57:43.5			69	148	339	6	0.855	0.796

Table 10b
LOCAL CIRCUMSTANCES DURING THE ANNULAR SOLAR ECLIPSE OF 10 MAY 1994
FOR THE UNITED STATES OF AMERICA

Location Name	First Contact U.T. h m s	Alt	P	V	Second Contact U.T. h m s	Alt	P	V	Third Contact U.T. h m s	Alt	P	V	Fourth Contact U.T. h m s	Alt	P	V
OHIO																
Kettering	15:25:55.4	55	254	299									18:57:43.6	61	73	37
Lakewood	15:33:29.3	57	253	292	17:14:24.2	66	291	295	17:19:21.8	66	36	38	19:03:21.9	58	75	38
Lima	15:28:04.7	55	252	295	17:08:28.0	66	299	310	17:12:49.1	67	27	37	18:58:12.8	61	74	40
Lorain	15:32:47.1	56	253	292	17:13:21.5	66	286	292	17:18:37.6	66	41	44	19:02:33.7	58	75	39
Mansfield	15:30:57.5	56	253	295									19:01:51.7	59	74	37
Parma	15:33:26.6	57	253	292	17:14:46.2	66	300	304	17:19:07.8	66	28	30	19:03:31.5	58	75	38
Springfield	15:27:06.1	55	254	298									18:58:45.1	61	73	37
Steubenville	15:33:55.4	58	255	295									19:06:05.4	58	73	33
Toledo	15:30:38.2	55	252	292	17:09:45.9	66	254	264	17:15:59.4	66	72	79	18:59:24.9	59	76	42
Warren	15:34:55.0	57	254	292									19:05:30.8	57	75	36
Youngstown	15:35:00.5	58	254	292									19:05:56.4	57	75	36
OKLAHOMA																
Clinton	14:57:18.5	39	250	308	16:26:51.5	57	263	314	16:32:39.4	58	55	106	18:16:43.2	72	70	80
Enid	15:00:11.3	41	249	306	16:30:31.1	58	247	296	16:36:30.7	59	72	119	18:20:42.2	71	71	75
Lawton	14:56:07.5	40	251	311									18:17:01.4	73	68	77
Midwest City	14:58:50.6	41	251	309	16:31:30.3	59	316	5	16:33:53.7	59	2	52	18:20:51.8	72	69	72
Norman	14:58:23.4	41	251	309									18:20:30.2	72	69	73
Muskogee	15:01:47.7	43	252	309									18:26:39.3	72	69	63
Oklahoma City	14:58:46.0	41	250	309	16:30:46.9	59	305	354	16:34:12.5	59	14	63	18:20:31.9	72	69	73
Ponca City	15:01:39.4	42	249	306	16:32:37.0	59	250	297	16:38:38.2	60	69	115	18:23:07.6	71	71	72
Tulsa	15:01:51.7	43	251	308	16:35:31.7	60	312	359	16:38:17.6	61	7	52	18:25:32.4	72	70	66
OREGON																
Burns	15:05:15.5	25	230	279									17:39:39.4	52	93	133
Corvallis	15:07:31.3	23	226	275									17:33:03.7	48	97	139
Eugene	15:06:15.3	22	227	276									17:32:47.7	48	96	139
Medford	15:02:05.2	22	229	280									17:31:16.6	49	93	139
Pendleton	15:10:25.6	26	228	275									17:42:17.6	51	95	132
Portland	15:09:51.9	23	226	273									17:35:08.2	48	97	138
Salem	15:08:24.6	23	226	274									17:33:51.5	48	97	138
PENNSYLVANIA																
Allentown	15:44:58.7	62	258	289									19:16:48.9	53	74	28
Altoona	15:38:27.8	60	256	293									19:10:51.3	55	73	31
Bethlehem	15:45:16.2	62	257	288									19:16:57.2	52	74	29
Erie	15:37:45.8	58	253	289	17:18:39.5	66	278	277	17:24:23.6	66	51	48	19:06:56.2	56	76	38
Harrisburg	15:41:32.1	61	257	291									19:14:06.2	54	73	29
Lancaster	15:42:29.3	62	258	291									19:15:16.4	54	73	28
Penn Hills	15:35:36.3	59	255	294									19:07:45.6	57	74	32
Philadelphia	15:45:02.6	63	259	290									19:17:41.7	52	73	26
Pittsburgh	15:35:19.3	59	255	294									19:07:30.3	57	74	33
Reading	15:43:44.1	62	258	290									19:16:03.4	53	73	28
Scranton	15:45:40.5	62	256	286									19:16:10.4	52	75	31
Upper Darby	15:44:44.3	63	259	290									19:17:28.1	53	73	26
Wilkes-Barre	15:44:58.9	61	256	287									19:15:48.5	53	75	31
RHODE ISLAND																
Cranston	15:55:36.4	64	258	278									19:23:54.0	48	76	30
East Providen..	15:55:46.4	64	258	278									19:23:57.3	48	76	30
Pawtucket	15:55:48.3	64	258	278									19:23:53.0	48	76	30
Providence	15:55:37.9	64	258	278									19:23:50.6	48	76	30
Warwick	15:55:27.6	64	258	278									19:23:53.0	48	76	29
SOUTH CAROLINA																
Charleston	15:25:15.2	61	266	320									19:04:38.9	61	60	7
Columbia	15:24:08.1	59	263	316									19:03:01.6	62	62	12
Greenville	15:22:06.4	58	261	314									19:00:08.3	63	64	18
North Charles..	15:25:17.1	61	266	320									19:04:40.8	61	60	7
Spartanburg	15:23:13.8	58	262	313									19:01:22.0	62	64	17
SOUTH DAKOTA																
Pierre	15:15:18.1	40	239	285									18:20:49.4	63	84	91
Rapid City	15:12:25.3	38	238	285									18:13:58.3	63	85	99
Sioux Falls	15:16:39.4	43	242	288									18:28:59.3	64	82	79
TENNESSEE																
Chattanooga	15:16:27.0	54	259	313									18:52:49.9	66	65	24
Clarksville	15:14:57.2	52	256	309									18:48:30.0	67	68	34
Knoxville	15:20:31.0	56	259	310									18:56:58.7	64	66	24
Memphis	15:08:14.6	49	256	313									18:40:23.8	70	66	38
Nashville	15:15:27.5	53	257	310									18:49:50.4	66	67	31

Table 10a
CIRCUMSTANCES AT MAXIMUM ECLIPSE ON 10 MAY 1994
FOR THE UNITED STATES OF AMERICA

Location Name	Latitude ° ′	Longitude ° ′	Elev. m	U.T. h m s	Umbral Durat. m s	Path Width km	Sun Alt °	Sun Az. °	P °	V °	Eclipse Mag.	Eclipse Obs.
TEXAS												
Abilene	32 25.0	-99-45.0	561	16:22:12.4			57	108	337	35	0.902	0.851
Amarillo	35 12.0	-101-51.0	1209	16:23:59.0	4 36.3	247	54	110	158	211	0.942	0.887
Arlington	32 44.0	-97-07.0	–	16:27:54.9			60	113	337	32	0.879	0.824
Austin	30 17.4	-97-43.8	196	16:21:49.2			59	107	337	37	0.828	0.764
Baytown	29 44.0	-95-01.0	–	16:26:20.6			62	109	337	36	0.782	0.709
Beaumont	30 04.8	-94-07.2	7	16:29:00.8			63	112	337	35	0.780	0.707
Brownsville	25 54.6	-97-29.4	5	16:13:25.1			58	98	336	45	0.721	0.636
Corpus Christi	27 45.0	-97-24.6	11	16:17:19.4			59	102	336	42	0.764	0.687
Dallas	32 47.4	-96-47.4	143	16:28:41.1			60	113	337	32	0.876	0.821
El Paso	31 47.4	-106-25.2	1285	16:09:55.6	5 39.7	256	49	99	337	39	0.941	0.885
Fort Worth	32 44.9	-97-19.7	220	16:27:31.2			59	112	337	32	0.881	0.827
Galveston	29 18.0	-94-48.6	2	16:25:55.5			62	109	337	37	0.769	0.694
Garland	32 55.0	-96-39.0	–	16:29:13.1			60	114	337	31	0.878	0.823
Grand Prairie	32 45.0	-97 00.0	–	16:28:11.0			60	113	337	32	0.878	0.823
Houston	29 45.0	-95-23.4	13	16:25:34.2			62	109	337	37	0.787	0.715
Irving	32 49.0	-96-57.0	–	16:28:24.9			60	113	337	32	0.879	0.824
Laredo	27 31.0	-99-29.0	144	16:12:50.0			56	99	336	43	0.785	0.712
Longview	32 29.0	-94-44.0	–	16:32:23.6			63	116	337	30	0.845	0.785
Lubbock	33 35.0	-101-51.0	1048	16:20:47.0	4 37.6	248	54	107	337	34	0.942	0.887
McAllen	26 12.0	-98-13.0	–	16:12:34.5			57	98	336	45	0.737	0.655
Mesquite	32 46.0	-96-35.0	–	16:29:03.6			60	114	337	32	0.873	0.818
Midland	32 05.0	-102-05.0	–	16:17:23.8			54	105	337	37	0.922	0.871
Odessa	31 51.0	-102-22.0	–	16:16:26.8			53	104	337	37	0.920	0.869
Pasadena	29 43.0	-95-13.0	–	16:25:52.6			62	109	337	36	0.784	0.712
Plano	33 01.0	-96-42.0	–	16:29:18.8			60	114	337	31	0.880	0.826
Port Arthur	29 52.0	-93-59.0	3	16:28:53.7			63	111	337	35	0.773	0.698
Plainsview	34 11.0	-101-43.0	–	16:22:12.8	5 44.7	248	55	109	337	33	0.942	0.887
Richardson	32 56.0	-96-44.0	–	16:29:04.9			60	114	337	31	0.879	0.824
San Angelo	31 28.0	-100-22.0	605	16:19:10.6			56	105	337	37	0.887	0.834
San Antonio	29 25.8	-98-30.0	213	16:18:34.5			58	104	336	39	0.817	0.751
Tyler	32 21.0	-95-19.0	–	16:30:52.9			62	115	337	31	0.849	0.789
Victoria	28 48.0	-97 00.0	–	16:20:17.0			60	105	336	40	0.784	0.711
Waco	31 33.2	-97-08.0	133	16:25:32.6			60	110	337	34	0.851	0.791
Wichita Falls	33 54.0	-98-30.0	310	16:27:29.8			58	113	337	31	0.922	0.872
UTAH												
Logan	41 46.0	-111-51.0	–	16:22:24.0			45	108	159	208	0.710	0.622
Ogden	41 13.5	-111-58.4	1409	16:21:11.1			44	107	159	208	0.719	0.633
Orem	40 15.0	-111-50.0	–	16:19:27.3			45	106	159	210	0.740	0.657
Provo	40 15.0	-111-40.0	1493	16:19:38.8			45	106	159	210	0.741	0.659
Salt Lake City	40 45.6	-111-52.2	1385	16:20:24.0			45	107	159	209	0.729	0.645
Sandy City	40 36.0	-111-53.0	–	16:20:04.8			44	106	159	209	0.732	0.649
VERMONT												
Brattleboro	42 51.1	-72-33.8	98	17:39:35.8	2 8.3	231	62	208	345	324	0.943	0.889
Burlington	44 28.8	-73-13.2	36	17:38:32.7	4 26.5	230	61	205	165	146	0.943	0.889
Montpelier	44 15.6	-72-34.2	159	17:39:55.7	5 36.6	231	61	207	165	145	0.943	0.889
VIRGINIA												
Alexandria	38 49.2	-77-04.8	–	17:26:15.8			68	194	343	331	0.854	0.796
Arlington	38 55.0	-77-10.0	–	17:26:06.6			68	194	343	332	0.857	0.799
Bristol	36 36.6	-82-10.8	–	17:10:19.6			71	169	341	350	0.828	0.764
Charlottesville	38 02.4	-78-29.4	–	17:21:50.3			70	188	342	336	0.841	0.780
Chesapeake	38 48.0	-76-16.0	–	17:28:24.0			68	197	343	329	0.849	0.790
Danville	36 35.4	-79-24.0	–	17:17:53.9			71	183	342	339	0.807	0.739
Hampton	37 02.0	-76-21.0	–	17:26:49.1			70	198	343	328	0.801	0.731
Lynchburg	37 24.6	-79-09.6	–	17:19:24.9			70	185	342	338	0.828	0.764
Newport News	37 03.0	-76-28.8	–	17:26:28.2			70	197	343	328	0.802	0.733
Norfolk	36 54.0	-76-16.2	3	17:26:56.1			70	198	343	328	0.796	0.727
Petersburg	37 13.2	-77-24.0	–	17:24:03.6			70	193	342	332	0.812	0.745
Portsmouth	36 50.0	-76-19.0	3	17:26:45.0			70	198	343	328	0.795	0.725
Richmond	37 32.4	-77-27.6	52	17:24:10.6			70	192	343	332	0.821	0.756
Roanoke	37 16.8	-79-57.6	297	17:17:06.5			70	181	342	341	0.830	0.767
Virginia Beach	36 50.0	-75-58.0	–	17:27:44.0			70	199	343	327	0.793	0.722

Table 10b

LOCAL CIRCUMSTANCES DURING THE ANNULAR SOLAR ECLIPSE OF 10 MAY 1994 FOR THE UNITED STATES OF AMERICA

Location Name	First Contact				Second Contact				Third Contact				Fourth Contact			
	U.T. h m s	Alt °	P °	V °	U.T. h m s	Alt °	P °	V °	U.T. h m s	Alt °	P °	V °	U.T. h m s	Alt °	P °	V °
TEXAS																
Abilene	14:50:41.1	38	253	315									18:10:11.4	74	65	86
Amarillo	14:53:59.6	36	248	307	16:21:38.4	54	211	265	16:26:14.7	55	107	161	18:08:47.6	71	71	93
Arlington	14:54:08.8	41	254	316									18:17:49.8	75	65	70
Austin	14:49:15.4	39	257	322									18:11:59.0	77	61	76
Baytown	14:51:52.5	42	259	325									18:18:29.0	78	59	56
Beaumont	14:53:41.0	43	259	325									18:21:39.6	77	59	50
Brownsville	14:43:05.1	38	262	333									18:03:16.5	80	53	86
Corpus Christi	14:45:43.9	39	260	328									18:07:44.1	79	56	79
Dallas	14:54:38.0	41	254	316									18:18:48.7	75	65	68
El Paso	14:44:02.0	31	249	312	16:07:06.4	48	256	318	16:12:46.0	49	60	122	17:51:52.3	69	68	114
Fort Worth	14:53:55.4	40	254	316									18:17:16.5	75	65	72
Galveston	14:51:30.6	42	260	326									18:18:13.5	78	58	55
Garland	14:55:01.7	41	254	316									18:19:23.2	75	65	68
Grand Prairie	14:54:18.9	41	254	316									18:18:10.5	75	65	70
Houston	14:51:23.1	42	259	325									18:17:28.0	78	59	59
Irving	14:54:29.5	41	254	316									18:18:24.8	75	65	69
Laredo	14:42:55.1	36	259	328									18:01:34.0	78	57	94
Longview	14:56:43.1	43	256	318									18:24:00.2	75	63	56
Lubbock	14:50:51.3	36	250	311	16:18:30.6	54	287	344	16:23:08.2	55	30	87	18:06:27.6	72	68	95
McAllen	14:42:32.9	37	262	332									18:01:59.4	79	54	90
Mesquite	14:54:50.6	41	254	316									18:19:20.4	75	64	67
Midland	14:47:52.7	35	252	315									18:03:26.7	73	66	97
Odessa	14:47:12.5	35	252	315									18:02:18.4	73	66	99
Pasadena	14:51:34.3	42	259	325									18:17:53.4	78	59	58
Plano	14:55:08.8	41	254	315									18:19:24.1	75	65	68
Port Arthur	14:53:33.5	43	260	325									18:21:38.9	78	58	49
Plainsview	14:52:08.0	36	249	310	16:19:21.4	54	261	317	16:25:06.2	55	56	112	18:07:42.4	72	69	94
Richardson	14:54:57.5	41	254	316									18:19:11.1	75	65	68
San Angelo	14:48:22.6	37	253	317									18:06:56.4	75	64	90
San Antonio	14:46:58.2	38	257	323									18:08:13.5	77	60	83
Tyler	14:55:43.0	42	256	318									18:22:11.0	75	63	59
Victoria	14:47:48.6	39	259	326									18:11:05.5	78	58	73
Waco	14:52:04.5	40	255	319									18:15:52.2	76	63	71
Wichita Falls	14:54:42.8	39	252	312									18:15:49.4	73	67	78
UTAH																
Logan	15:02:29.9	30	235	287									17:52:33.8	59	86	120
Ogden	15:01:10.8	30	236	288									17:51:42.9	59	85	121
Orem	14:58:59.1	29	237	290									17:50:57.9	60	84	121
Provo	14:59:02.1	29	237	290									17:51:19.3	60	84	121
Salt Lake City	15:00:08.2	30	236	289									17:51:26.4	60	85	121
Sandy City	14:59:46.3	29	236	289									17:51:14.7	60	84	121
VERMONT																
Brattleboro	15:54:14.1	62	256	277	17:38:36.7	63	327	305	17:40:44.9	62	7	345	19:21:09.8	49	78	34
Burlington	15:54:51.6	61	253	274	17:36:16.1	61	213	195	17:40:42.6	61	120	101	19:18:47.2	49	80	39
Montpelier	15:55:55.5	61	254	273	17:37:05.3	61	233	214	17:42:41.9	61	101	80	19:20:05.2	48	80	38
VIRGINIA																
Alexandria	15:39:11.5	62	259	296									19:13:53.6	55	71	24
Arlington	15:39:07.2	62	259	296									19:13:42.1	55	71	25
Bristol	15:25:02.6	57	259	308									19:01:42.7	62	67	23
Charlottesvil...	15:35:00.5	61	259	301									19:10:47.1	57	69	23
Chesapeake	15:41:01.4	63	260	296									19:15:37.5	54	71	23
Danville	15:31:03.6	60	261	306									19:08:22.6	59	67	19
Hampton	15:38:45.6	63	262	302									19:15:25.0	55	68	18
Lynchburg	15:32:40.6	60	260	303									19:09:09.7	58	68	22
Newport News	15:38:27.8	63	262	302									19:15:08.0	55	68	18
Norfolk	15:38:48.4	64	262	302									19:15:34.5	55	68	17
Petersburg	15:36:28.1	62	261	302									19:13:07.0	56	68	19
Portsmouth	15:38:37.1	64	262	302									19:15:27.7	55	67	17
Richmond	15:36:43.0	62	261	301									19:13:01.7	56	69	20
Roanoke	15:30:43.9	60	260	304									19:07:16.8	59	68	22
Virginia Beach	15:39:28.6	64	263	302									19:16:13.9	55	67	17

CIRCUMSTANCES AT MAXIMUM ECLIPSE ON 10 MAY 1994
FOR THE UNITED STATES OF AMERICA

Location Name	Latitude ° '	Longitude ° '	Elev. m	U.T. h m s	Umbral Durat. m s	Path Width km	Sun Alt °	Sun Az. °	P °	V °	Eclipse Mag.	Eclipse Obs.
WASHINGTON												
Bellevue	47 37.0	-122-12.0	–	16:23:26.7			36	104	161	205	0.511	0.396
Billingham	48 45.0	-122-28.6	–	16:25:25.4			36	105	161	203	0.491	0.374
Everett	47 59.0	-122-11.0	–	16:24:10.3			36	104	161	204	0.506	0.389
Mt. Rainier	46 50.0	-121-45.0	–	16:22:17.4			36	104	161	205	0.528	0.414
Olympia	47 03.0	-122-53.0	–	16:21:47.7			35	103	161	205	0.515	0.399
Pullman	46 46.0	-117-09.0	–	16:26:19.0			40	108	161	204	0.569	0.459
Richland	46 17.0	-119-17.0	–	16:23:21.8			38	105	161	205	0.559	0.448
Seattle	47 37.8	-122-19.8	131	16:?3:21.9			36	104	161	205	0.510	0.394
Spokane	47 40.2	-117-24.6	773	16:27:47.9			40	109	161	203	0.551	0.439
Tacoma	47 16.0	-122-30.0	36	16:22:31.3			36	103	161	205	0.515	0.399
Walla Walla	46 05.0	-118-18.0	–	16:23:53.2			39	106	160	205	0.571	0.461
Yakima	46 35.7	-120-30.8	348	16:22:52.4			37	105	161	205	0.543	0.430
WEST VIRGINIA												
Charleston	38 21.0	-81-37.8	197	17:13:52.8			69	174	341	346	0.870	0.815
Greenbank	38 26.3	-79-50.2	–	17:18:39.1			69	182	342	340	0.860	0.803
Huntington	38 24.6	-82-25.8	185	17:11:53.5			69	171	341	349	0.878	0.824
Wheeling	40 04.2	-80-42.0	213	17:18:05.1			68	179	342	343	0.910	0.860
WISCONSIN												
Appleton	44 14.0	-88-27.0	–	17:05:18.7			62	157	161	178	0.872	0.816
Eau Claire	44 48.6	-91-30.0	–	17:00:02.4			60	149	161	183	0.834	0.772
Green Bay	44 30.0	-88-04.0	194	17:06:25.7			62	158	161	177	0.868	0.812
Janesville	42 41.0	-89-03.0	–	17:02:01.2			63	153	161	181	0.905	0.855
Kenosha	42 34.0	-87-50.0	–	17:04:26.8			63	156	161	179	0.918	0.868
La Crosse	43 48.6	-91-13.8	–	16:59:09.4			61	148	161	184	0.860	0.802
Madison	43 05.4	-89-23.4	282	17:01:52.2			62	153	161	182	0.892	0.840
Milwaukee	43 03.0	-87-57.0	208	17:04:49.8			63	157	161	179	0.905	0.854
Oshkosh	44 01.0	-88-35.0	–	17:04:45.7			62	156	161	179	0.876	0.821
Racine	42 43.0	-87-49.0	207	17:04:40.7			63	157	161	179	0.914	0.864
Sheboygan	43 45.6	-87-44.9	207	17:06:09.6			63	158	161	178	0.889	0.836
Waukesha	43 01.0	-88-13.0	–	17:04:13.3			63	156	161	179	0.903	0.853
Wauwatosa	43 03.0	-88 00.0	–	17:04:43.4			63	157	161	179	0.904	0.854
West Allis	43 01.0	-88-01.0	–	17:04:38.7			63	156	161	179	0.905	0.855
WYOMING												
Casper	42 50.4	-106-19.2	–	16:31:43.3			50	117	159	203	0.743	0.662
Cheyenne	41 08.4	-104-48.0	2010	16:30:43.3			51	116	159	204	0.793	0.721
Sheridan	44 47.8	-106-57.7	1301	16:34:28.5			49	119	160	201	0.698	0.608

LOCAL CIRCUMSTANCES DURING THE ANNULAR SOLAR ECLIPSE OF 10 MAY 1994
FOR THE UNITED STATES OF AMERICA

Location Name	First Contact U.T. h m s	Alt °	P °	V °	Second Contact U.T. h m s	Alt °	P °	V °	Third Contact U.T. h m s	Alt °	P °	V °	Fourth Contact U.T. h m s	Alt °	P °	V °
WASHINGTON																
Bellevue	15:15:09.4	25	224	269									17:37:58.8	48	100	137
Billingham	15:18:04.3	25	223	267									17:38:52.6	47	101	137
Everett	15:16:05.8	25	224	269									17:38:21.0	48	100	137
Mt. Rainier	15:13:10.1	25	225	271									17:38:01.5	48	99	136
Olympia	15:13:42.8	24	224	270									17:36:15.3	47	99	138
Pullman	15:13:24.5	28	227	273									17:46:25.7	52	96	129
Richland	15:11:55.0	26	227	273									17:41:58.3	51	96	133
Seattle	15:15:11.4	25	224	269									17:37:45.9	48	100	137
Spokane	15:15:40.2	28	226	271									17:46:41.7	51	97	130
Tacoma	15:14:15.7	24	224	270									17:37:07.6	48	100	137
Walla Walla	15:11:31.2	27	227	274									17:43:47.6	51	96	131
Yakima	15:12:36.7	25	226	272									17:40:00.3	49	98	135
WEST VIRGINIA																
Charleston	15:28:46.3	58	257	302									19:03:38.0	60	70	28
Greenbank	15:32:37.4	59	258	300									19:07:47.3	58	70	26
Huntington	15:27:16.2	57	256	302									19:01:45.6	61	70	30
Wheeling	15:33:17.6	58	255	296									19:05:55.2	58	73	32
WISCONSIN																
Appleton	15:27:59.0	50	246	286									18:48:14.3	61	81	59
Eau Claire	15:25:14.9	48	243	285									18:41:20.2	62	82	67
Green Bay	15:29:03.9	50	246	286									18:49:04.0	60	81	59
Janesville	15:23:58.0	50	247	291									18:46:51.4	62	78	56
Kenosha	15:25:29.0	51	248	291									18:49:10.4	62	78	53
La Crosse	15:23:25.3	48	245	288									18:41:54.1	63	81	64
Madison	15:24:19.5	49	246	290									18:46:06.2	62	79	58
Milwaukee	15:26:16.7	51	247	290									18:49:25.2	62	79	55
Oshkosh	15:27:21.1	50	246	287									18:47:57.2	61	80	59
Racine	15:25:48.3	51	248	291									18:49:43.2	62	78	53
Sheboygan	15:28:00.1	51	247	288									18:49:51.2	61	80	56
Waukesha	15:25:49.6	51	247	290									18:48:48.3	62	79	55
Wauwatosa	15:26:12.4	51	247	290									18:49:18.2	62	79	55
West Allis	15:26:06.9	51	247	290									18:49:15.9	62	79	55
WYOMING																
Casper	15:07:33.9	35	237	287									18:05:58.7	62	85	107
Cheyenne	15:04:31.1	36	240	292									18:08:04.7	64	81	103
Sheridan	15:11:54.5	35	235	282									18:05:59.4	60	88	109

Table 11a
CIRCUMSTANCES AT MAXIMUM ECLIPSE ON 10 MAY 1994
FOR CANADA

Location Name	Latitude ° ′	Longitude ° ′	Elev. m	U.T. h m s	Umbral Durat. m s	Path Width km	Sun Alt °	Sun Az. °	P °	V °	Eclipse Mag.	Eclipse Obs.
ALBERTA												
Banff	51 10.0	-115-34.0	—	16:36:13.0			41	116	161	198	0.508	0.392
Calgary	51 03.0	-114-05.0	1161	16:37:29.6			42	118	161	197	0.521	0.406
Edmonton	53 33.0	-113-28.0	728	16:42:38.1			42	121	162	194	0.484	0.366
Lethbridge	49 41.4	-112-49.2	979	16:36:20.1			43	117	161	198	0.555	0.444
Medicine Hat	50 03.0	-110-40.0	—	16:39:24.5			45	121	161	197	0.566	0.457
Red Deer	52 16.0	-113-48.0	—	16:40:00.2			42	119	162	196	0.503	0.387
BRITISH COLUMBIA												
Fort Nelson	58 50.0	-122-35.0	411	16:44:22.9			35	116	163	193	0.342	0.223
Kamloops	50 40.0	-120-20.0	—	16:30:53.2			38	109	161	200	0.478	0.360
Kelowna	49 53.0	-119-29.0	—	16:30:07.7			38	109	161	201	0.497	0.380
Matsqui	49 12.0	-122-25.0	—	16:26:20.6			36	105	161	203	0.484	0.367
Nanaimo	49 10.0	-123-56.0	—	16:25:06.0			35	104	161	203	0.473	0.354
Penticton	49 19.2	-119-37.2	370	16:28:55.6			38	109	161	202	0.505	0.389
Prince George	53 55.0	-122-45.0	728	16:35:07.1			36	110	162	198	0.410	0.290
Prince Rupert	54 19.2	-130-19.2	56	16:30:53.7			31	103	163	199	0.351	0.232
Vancouver	49 16.0	-123-07.0	42	16:25:55.2			35	105	161	203	0.478	0.359
Victoria	48 26.0	-123-23.0	19	16:24:05.7			35	104	161	204	0.489	0.371
MANITOBA												
Brandon	49 51.0	-99-57.0	415	16:53:09.4			52	138	161	188	0.656	0.559
Churchill	58 46.0	-94-10.0	31	17:12:19.2			47	158	164	176	0.512	0.397
Selkirk	50 10.0	-96-52.0	—	16:58:16.7			53	144	162	185	0.673	0.579
The Pas	53 49.8	-101-15.0	292	16:57:15.8			49	141	162	185	0.567	0.457
Winnipeg	49 53.0	-97-09.0	257	16:57:26.3			53	144	161	185	0.677	0.583
NEW BRUNSWICK												
Chatham	47 02.0	-65-28.0	36	17:54:20.1			55	222	168	139	0.900	0.848
Edmunston	47 22.0	-68-20.0	—	17:48:46.9			56	215	167	142	0.887	0.833
Fredericton	45 57.0	-66-38.5	10	17:52:22.6			56	221	167	139	0.927	0.877
Moncton	46 05.4	-64-47.4	12	17:55:57.4			55	225	168	137	0.926	0.876
St. John	45 16.0	-66-03.0	39	17:53:44.0	3 17.8	234	56	223	167	137	0.942	0.887
St. Stephen	45 12.0	-67-17.0	—	17:51:14.4	3 18.4	233	57	220	167	138	0.942	0.888
NEWFOUNDLAND												
Gander	48 57.0	-54-37.0	163	18:11:10.2			46	239	170	134	0.845	0.784
St. John's	47 34.0	-52-43.0	69	18:15:14.5			45	244	171	131	0.878	0.822
NOVA SCOTIA												
Bridgewater	44 23.0	-64-31.0	—	17:57:06.6	6 0.1	235	56	227	348	314	0.942	0.887
Dartmouth	44 40.0	-63-34.0	8	17:58:54.3	5 51.9	235	55	229	168	134	0.942	0.887
Glace Bay	46 12.0	-59-57.0	—	18:04:50.9			51	234	169	133	0.924	0.873
Guysborough	45 23.0	-61-30.0	—	18:02:31.8	3 10.3	237	53	232	169	133	0.942	0.886
Halifax	44 39.0	-63-36.0	27	17:58:50.8	5 53.0	235	55	229	168	134	0.942	0.887
Kentville	45 05.0	-64-30.0	—	17:56:53.5	4 45.2	235	55	226	168	135	0.942	0.887
Liverpool	44 02.0	-64-43.0	—	17:56:49.2	5 34.8	235	56	227	348	314	0.942	0.887
New Glasgow	45 35.0	-62-39.0	—	18:00:15.8			54	229	168	134	0.941	0.886
Port Hawkesbury	45 37.0	-61-21.0	—	18:02:40.6			52	232	169	133	0.940	0.886
Sable Island	43 55.0	-59-50.0	—	18:06:32.0	5 5.9	238	52	237	349	310	0.941	0.886
Shelburne	43 46.0	-65-19.0	—	17:55:39.4	4 46.9	234	57	226	348	314	0.942	0.888
Sydney	46 08.4	-60-10.8	5	18:04:28.8			51	233	169	133	0.925	0.875
Truro	45 22.0	-63-16.0	—	17:59:11.2	3 11.7	235	54	229	168	135	0.942	0.887
Windsor	44 59.0	-64-08.0	—	17:57:39.3	5 9.9	235	55	227	168	135	0.942	0.887
Yarmouth	43 50.0	-66-07.0	—	17:53:57.3	5 11.3	234	57	224	347	315	0.942	0.888
NORTHWEST TERRITORY												
Aklavik	68 14.0	-135 00.0	10	16:55:01.9			27	111	165	187	0.175	0.084
Alert	82 31.0	-62-20.0	31	17:36:22.9			25	204	169	166	0.154	0.070
Fort Simpson	61 45.0	-121-14.0	182	16:50:31.3			35	121	164	189	0.312	0.195
Frobisher Bay	63 45.0	-68-33.0	36	17:43:35.9			42	204	168	157	0.486	0.369
Resolute	74 43.0	-94-59.0	72	17:23:58.9			33	165	167	171	0.234	0.129
Yellowknife	62 27.0	-114-21.0	221	16:56:46.4			38	130	164	186	0.340	0.222

66

LOCAL CIRCUMSTANCES DURING THE ANNULAR SOLAR ECLIPSE OF 10 MAY 1994
FOR CANADA

Location Name	First Contact				Second Contact				Third Contact				Fourth Contact			
	U.T. h m s	Alt	P	V	U.T. h m s	Alt	P	V	U.T. h m s	Alt	P	V	U.T. h m s	Alt	P	V
ALBERTA																
Banff	15:25:07.9	30	224	265									17:52:39.6	50	101	127
Calgary	15:25:09.5	31	225	265									17:55:18.8	51	100	125
Edmonton	15:32:00.1	32	223	260									17:57:40.8	50	103	125
Lethbridge	15:21:55.3	32	227	269									17:56:55.4	53	98	122
Medicine Hat	15:23:33.8	33	228	269									18:01:19.4	54	97	119
Red Deer	15:28:28.0	32	224	263									17:56:29.7	51	102	125
BRITISH COLUMBIA																
Fort Nelson	15:45:23.3	28	214	246									17:45:37.4	42	114	138
Kamloops	15:23:10.1	27	222	264									17:43:50.1	48	103	134
Kelowna	15:21:10.1	27	223	266									17:44:42.5	49	101	133
Matsqui	15:19:14.5	25	222	266									17:39:02.7	47	102	137
Nanaimo	15:19:10.6	24	222	265									17:36:28.1	46	103	139
Penticton	15:19:40.9	27	224	267									17:44:00.7	49	101	133
Prince George	15:31:48.7	27	218	256									17:42:15.1	44	108	137
Prince Rupert	15:33:33.5	22	214	252									17:31:21.9	39	113	146
Vancouver	15:19:25.0	25	222	265									17:37:55.2	46	102	138
Victoria	15:17:15.7	24	223	267									17:36:42.8	46	102	138
MANITOBA																
Brandon	15:28:55.1	41	233	272									18:23:01.1	58	92	96
Churchill	15:55:32.3	42	226	250									18:30:11.6	49	104	101
Selkirk	15:32:09.8	43	234	271									18:29:21.1	58	92	90
The Pas	15:38:08.1	40	228	262									18:20:13.1	54	99	104
Winnipeg	15:31:13.8	43	235	272									18:28:48.3	58	92	90
NEW BRUNSWICK																
Chatham	16:13:37.9	61	253	254									19:27:49.4	41	85	43
Edmunston	16:08:10.3	60	251	258									19:23:29.4	44	85	45
Fredericton	16:10:14.1	62	254	258									19:27:34.5	43	84	40
Moncton	16:14:16.7	62	254	255									19:29:54.4	41	84	40
St. John	16:10:55.0	62	255	259	17:52:01.4	56	202	172	17:55:19.2	56	136	105	19:29:11.4	42	83	38
St. Stephen	16:08:11.4	62	255	261	17:49:31.4	57	202	174	17:52:49.9	57	135	106	19:27:31.0	43	82	38
NEWFOUNDLAND																
Gander	16:36:21.2	56	252	234									19:36:28.5	33	91	48
St. John's	16:39:53.8	57	255	232									19:40:16.1	31	89	44
NOVA SCOTIA																
Bridgewater	16:13:39.7	63	257	257	17:54:06.7	56	262	229	18:00:06.8	55	76	42	19:32:16.8	41	81	35
Dartmouth	16:15:58.7	63	257	255	17:55:57.4	55	248	215	18:01:49.3	54	91	56	19:33:12.7	40	82	36
Glace Bay	16:24:41.7	61	255	245									19:35:34.7	37	85	40
Guysborough	16:20:57.7	62	256	249	18:00:53.2	53	203	167	18:04:03.5	53	138	102	19:34:53.8	38	84	38
Halifax	16:15:53.6	63	257	255	17:55:53.4	55	249	216	18:01:46.4	54	90	55	19:33:11.3	40	82	36
Kentville	16:14:10.3	63	256	256	17:54:28.2	56	222	190	17:59:13.4	55	117	84	19:31:30.2	41	83	37
Liverpool	16:12:58.7	64	257	258	17:54:03.5	56	282	248	17:59:38.4	56	57	23	19:32:23.5	41	81	34
New Glasgow	16:18:32.9	62	256	251									19:33:15.3	39	84	38
Port Hawkesbu..	16:21:24.3	62	256	249									19:34:46.3	38	84	38
Sable Island	16:24:09.7	63	259	248	18:04:01.1	52	291	252	18:09:07.0	52	50	10	19:38:39.3	37	82	34
Shelburne	16:11:25.6	64	258	260	17:53:18.9	57	297	264	17:58:05.7	56	41	7	19:31:50.6	42	80	34
Sydney	16:24:10.6	61	255	246									19:35:24.4	37	85	40
Truro	16:17:04.2	62	256	253	17:57:31.8	54	202	169	18:00:43.5	54	137	103	19:32:45.4	40	83	38
Windsor	16:14:54.8	63	256	255	17:55:02.1	55	229	197	18:00:12.1	55	110	76	19:32:06.2	41	82	37
Yarmouth	16:09:38.7	64	257	262	17:51:24.1	58	290	259	17:56:35.4	57	48	15	19:30:38.4	42	80	34
NORTHWEST TERRITORY																
Aklavik	16:13:09.6	24	201	224									17:36:56.6	31	131	150
Alert	16:57:15.4	25	203	201									18:14:17.1	24	137	133
Fort Simpson	15:53:36.5	29	212	241									17:48:55.0	40	117	137
Frobisher Bay	16:29:01.2	44	228	228									18:55:06.4	38	110	90
Resolute	16:33:45.6	31	208	216									18:13:10.3	33	128	128
Yellowknife	15:56:20.1	32	214	241									17:58:32.2	42	115	130

Location Name	Latitude ° '	Longitude ° '	Elev. m	U.T. h m s	Umbral Durat. m s	Path Width km	Sun Alt °	Sun Az. °	P °	V °	Eclipse Mag.	Eclipse Obs.
ONTARIO												
Barrie	44 24.0	-79-40.0	–	17:24:14.4			63	185	163	160	0.925	0.876
Brantford	43 08.0	-80-16.0	231	17:21:54.6	5 2.4	229	65	182	163	161	0.943	0.889
Cambridge	43 22.0	-80-19.0	–	17:21:59.2	3 40.1	229	64	182	163	161	0.943	0.889
Chatham	42 24.0	-82-11.0	–	17:16:50.8	5 53.0	230	65	175	162	166	0.943	0.889
Cornwall	45 02.0	-74-44.0	–	17:35:24.4			61	200	165	150	0.931	0.881
Guelph	43 33.0	-80-15.0	349	17:22:17.3	1 42.5	229	64	183	163	161	0.943	0.889
Hamilton	43 15.0	-79-51.0	108	17:22:57.3	4 55.9	229	64	184	163	160	0.943	0.889
Kapuskasing	49 25.0	-82-28.0	244	17:22:11.1			58	178	164	165	0.783	0.709
Kingston	44 15.0	-76-38.0	87	17:30:52.8	1 49.4	230	63	195	164	153	0.943	0.889
Kitchener	43 27.0	-80-29.0	361	17:21:40.8	2 28.2	229	64	182	163	161	0.943	0.889
London	42 59.0	-81-14.0	270	17:19:34.5	4 43.8	230	65	179	162	163	0.943	0.889
Niagara Falls	43 06.0	-79-04.0	194	17:24:38.1	5 53.4	229	64	187	163	158	0.943	0.889
North Bay	49 19.0	-79-28.0	399	17:27:43.4			58	186	164	160	0.800	0.730
Ottawa	45 25.0	-75-42.0	123	17:33:28.0			61	197	165	152	0.917	0.868
Oshawa	43 54.0	-78-51.0	115	17:25:42.1			64	187	163	158	0.943	0.889
Peterborough	44 18.0	-78-19.5	221	17:27:08.5			63	189	164	157	0.935	0.884
Port Arthur	48 22.0	-89-19.0	211	17:08:37.6			58	159	162	176	0.767	0.690
Pt. Pelee N.P.	41 57.0	-82-30.0	–	17:15:39.5	6 13.6	230	66	174	342	347	0.943	0.890
St. Catharines	43 10.0	-79-15.0	119	17:24:15.8	5 39.2	229	64	186	163	158	0.943	0.889
St. Thomas	42 47.0	-81-12.0	–	17:19:28.1	5 31.6	230	65	179	162	163	0.943	0.889
Sarnia	42 58.0	-82-23.0	–	17:16:56.8	2 55.1	230	65	175	162	166	0.943	0.889
Sault St. Marie	46 31.8	-84-19.8	193	17:16:12.7			61	171	163	169	0.844	0.783
Sudbury	46 28.0	-81 00.0	279	17:22:55.1			61	181	163	162	0.864	0.807
Thunder Bay	48 25.2	-89-13.8	202	17:08:50.6			58	160	162	176	0.766	0.690
Toronto	43 39.0	-79-23.0	124	17:24:19.3	2 54.4	229	64	186	163	159	0.943	0.889
Welland	43 04.0	-79-03.0	–	17:24:38.9	5 57.5	229	65	187	163	158	0.943	0.889
Windsor	42 18.0	-83-01.0	198	17:14:49.8	5 37.3	230	65	172	162	168	0.943	0.889
Woodstock	43 08.0	-80-45.0	–	17:20:48.4	4 32.4	229	65	181	163	162	0.943	0.889
Prince Edward Island												
Charlottetown	46 14.0	-63-08.0	59	17:59:02.5			54	228	168	136	0.923	0.873
QUEBEC												
Chicoutimi	48 26.0	-71-04.0	–	17:43:27.5			57	208	166	147	0.852	0.793
Drummondville	45 53.0	-72-29.0	–	17:40:25.6			60	206	166	147	0.916	0.867
Knob Lake	54 48.0	-66-49.0	562	17:49:16.4			49	212	168	149	0.697	0.607
Montreal	45 31.0	-73-34.0	61	17:38:03.8			60	203	165	148	0.922	0.873
Quebec City	46 49.0	-71-14.0	78	17:43:05.7			58	209	166	146	0.895	0.843
Shawinigan	46 33.0	-72-45.0	–	17:39:59.1			59	205	166	148	0.898	0.846
Sherbrooke	45 25.0	-71-54.0	176	17:41:35.2			60	208	166	145	0.931	0.881
Trois-Rivieres	46 21.0	-72-33.0	38	17:40:21.9			59	206	166	147	0.903	0.853
SASKATCHEWAN												
Moose Jaw	50 23.0	-105-32.0	585	16:46:16.6			48	129	161	193	0.602	0.496
Regina	50 25.0	-104-39.0	618	16:47:29.3			49	131	161	192	0.608	0.504
Saskatoon	52 07.0	-106-38.0	554	16:47:46.9			47	130	162	192	0.560	0.450
YUKON TERRITORY												
Inuvik	68 25.0	-133-30.0	–	16:55:56.2			28	113	165	186	0.180	0.088
Whitehorse	60 43.0	-135-03.0	756	16:40:51.8			28	104	164	194	0.246	0.139

68

LOCAL CIRCUMSTANCES DURING THE ANNULAR SOLAR ECLIPSE OF 10 MAY 1994
FOR CANADA

Location Name	First Contact U.T. h m s	Alt °	P °	V °	Second Contact U.T. h m s	Alt °	P °	V °	Third Contact U.T. h m s	Alt °	P °	V °	Fourth Contact U.T. h m s	Alt °	P °	V °
ONTARIO																
Barrie	15:42:18.0	57	250	281									19:06:51.0	55	80	45
Brantford	15:39:05.0	57	251	286	17:19:20.4	65	219	218	17:24:22.8	65	110	107	19:06:12.7	56	78	41
Cambridge	15:39:23.0	57	251	285	17:20:05.0	64	201	200	17:23:45.1	64	128	125	19:06:00.9	56	78	42
Chatham	15:34:21.8	56	251	289	17:13:52.6	65	235	240	17:19:45.6	65	92	95	19:02:22.2	58	77	42
Cornwall	15:52:34.5	60	252	274									19:15:41.6	50	81	42
Guelph	15:39:48.7	57	251	284	17:21:21.1	64	181	179	17:23:03.6	64	148	146	19:06:04.9	55	79	43
Hamilton	15:40:02.7	57	252	285	17:20:26.2	64	217	215	17:25:22.1	64	112	108	19:07:01.3	55	78	41
Kapuskasing	15:46:58.3	52	242	269									18:58:05.3	53	88	63
Kingston	15:47:43.6	59	252	279	17:29:53.3	63	183	172	17:31:42.6	63	148	137	19:12:48.0	52	80	41
Kitchener	15:39:13.2	57	251	285	17:20:21.9	64	188	187	17:22:50.2	64	141	139	19:05:38.6	56	79	43
London	15:37:04.3	56	251	287	17:17:09.2	65	214	216	17:21:53.0	65	114	114	19:04:14.9	57	78	42
Niagara Falls	15:41:16.2	58	252	285	17:21:39.8	65	236	233	17:27:33.2	64	93	87	19:08:40.7	55	78	40
North Bay	15:51:15.5	54	244	267									19:03:34.2	51	88	59
Ottawa	15:51:15.9	59	251	274									19:13:40.1	51	82	44
Oshawa	15:42:58.0	58	251	282									19:08:43.9	54	79	42
Peterborough	15:44:35.4	58	251	280									19:09:32.5	53	80	43
Port Arthur	15:35:42.5	48	240	275									18:45:22.0	58	87	71
Pt. Pelee N.P.	15:33:01.0	56	252	291	17:12:32.9	66	255	261	17:18:46.5	66	73	76	19:01:45.6	58	76	41
St. Catharines	15:41:01.8	58	252	285	17:21:24.1	64	230	227	17:27:03.3	64	99	93	19:08:16.7	55	78	40
St. Thomas	15:36:47.4	57	251	288	17:16:40.0	65	227	229	17:22:11.6	65	101	101	19:04:22.6	57	77	42
Sarnia	15:35:00.5	56	251	288	17:15:24.6	65	192	196	17:18:19.7	65	136	139	19:01:48.3	58	78	44
Sault St. Mar...	15:38:33.7	53	245	278									18:56:22.9	57	84	57
Sudbury	15:43:38.3	55	247	276									19:03:01.3	54	83	52
Thunder Bay	15:35:55.9	49	240	275									18:45:31.4	58	87	71
Toronto	15:41:34.1	58	251	283	17:22:47.5	64	193	189	17:25:41.9	64	137	132	19:07:47.5	55	79	42
Welland	15:41:14.9	58	252	285	17:21:38.7	65	239	235	17:27:36.2	64	91	85	19:08:43.5	55	78	40
Windsor	15:32:43.2	55	251	290	17:11:59.0	65	228	236	17:17:36.3	65	99	103	19:00:34.3	59	77	43
Woodstock	15:38:12.1	57	251	286	17:18:28.7	65	211	212	17:23:01.0	65	117	116	19:05:12.4	56	78	42
PRINCE EDWARD ISLAND																
Charlottetown	16:17:54.5	61	255	251									19:31:49.7	40	85	40
QUEBEC																
Chicoutimi	16:04:12.1	58	249	260									19:18:12.2	46	87	49
Drummondville	15:58:10.1	60	252	269									19:18:46.6	48	83	43
Knob Lake	16:19:21.6	53	241	242									19:14:12.5	41	97	66
Montreal	15:55:32.4	60	252	271									19:17:18.5	49	82	42
Quebec City	16:01:49.2	60	251	264									19:19:49.1	46	84	45
Shawinigan	15:58:31.8	59	250	267									19:17:40.9	48	84	45
Sherbrooke	15:58:44.6	61	252	269									19:20:10.9	47	82	41
Trois-Rivieres	15:58:39.0	60	251	267									19:18:12.6	48	83	44
SASKATCHEWAN																
Moose Jaw	15:26:43.2	37	230	270									18:11:40.3	56	95	109
Regina	15:27:17.4	37	230	270									18:13:27.8	56	95	107
Saskatoon	15:30:40.4	36	228	265									18:09:54.6	54	98	112
YUKON TERRITORY																
Inuvik	16:13:22.9	24	202	224									17:38:31.5	31	130	149
Whitehorse	15:52:09.7	22	207	238									17:30:50.7	34	123	150

Table 12a
CIRCUMSTANCES AT MAXIMUM ECLIPSE ON 10 MAY 1994
FOR EUROPE

Location Name	Latitude ° ′	Longitude ° ′	Elev. m	U.T. h m s	Umbral Durat. m s	Path Width km	Sun Alt °	Sun Az. °	P °	V °	Eclipse Mag.	Eclipse Obs.
ANDORRA												
Andorra la Vella	42 30.0	1 31.0	1162	18:46:57.7			2	292	177	131	0.700	0.608
AUSTRIA												
Vienna	48 13.0	16 20.0	218	18:22 Set			0	298	–	–	0.454	0.334
BELGIUM												
Antwerp	51 13.0	4 25.0	–	18:35:18.5			5	292	176	138	0.516	0.400
Brussels	50 50.0	4 20.0	–	18:35:50.2			5	292	176	138	0.523	0.408
Liege	50 38.0	5 34.0	–	18:35:44.3			4	293	176	138	0.520	0.404
BYELARUS												
Minsk	53 54.0	27 35.0	242	17:59 Set			0	303	–	–	0.235	0.129
CZECHOSLOVAKIA												
Ostrava	49 50.0	18 17.0	–	18:16 Set			0	299	–	–	0.398	0.276
Prague	50 05.0	14 28.0	217	18:33:03.7			0	299	176	140	0.477	0.358
DENMARK												
Copenhagen	55 40.0	12 35.0	14	18:26:53.6			4	296	175	143	0.398	0.277
ESTONIA												
Tallinn	59 26.0	24 44.0	–	18:17:18.7			2	304	175	148	0.286	0.172
FINLAND												
Helsinki	60 10.0	24 58.0	10	18:16:18.9			2	304	175	149	0.276	0.163
FRANCE												
Bordeaux	44 50.0	0-34.0	52	18:44:46.7			5	291	177	132	0.667	0.570
Lille	50 38.0	3 04.0	46	18:36:26.4			6	291	176	137	0.534	0.420
Lyon	45 43.0	5 04.0	308	18:41:59.6			2	294	177	134	0.613	0.507
Marseille	43 18.0	5 24.0	81	18:44:39.3			1	295	177	133	0.658	0.559
Paris	48 52.0	2 20.0	54	18:38:56.3			5	291	176	136	0.571	0.460
Toulouse	43 36.0	1 26.0	177	18:45:41.8			3	292	177	132	0.678	0.583
GERMANY												
Aachen	50 47.0	6 05.0	–	18:35:23.2			4	293	176	138	0.514	0.398
Berlin	52 31.0	13 24.0	61	18:30:35.8			2	298	176	141	0.443	0.322
Bielefeld	52 01.0	8 31.0	–	18:33:01.1			4	295	176	140	0.479	0.360
Bonn	50 44.0	7 05.0	–	18:35:08.0			4	294	176	138	0.509	0.392
Bremen	53 04.0	8 49.0	17	18:31:33.6			4	294	176	140	0.460	0.340
Dortmund	51 31.0	7 28.0	–	18:34:00.3			4	294	176	139	0.493	0.375
Dresden	51 03.0	13 44.0	–	18:32:14.4			1	298	176	140	0.465	0.345
Duisburg	51 25.0	6 46.0	–	18:34:21.4			4	294	176	139	0.499	0.381
Dusseldorf	51 12.0	6 47.0	–	18:34:37.9			4	294	176	139	0.503	0.385
Essen	52 43.0	7 57.0	–	18:32:17.6			4	294	176	140	0.470	0.351
Frankfurt	50 07.0	8 40.0	111	18:35:22.1			3	295	176	138	0.510	0.394
Hamburg	53 33.0	9 59.0	22	18:30:32.7			4	295	176	141	0.445	0.325
Hannover	52 24.0	9 44.0	–	18:32:06.8			3	295	176	140	0.466	0.346
Koln	50 56.0	6 59.0	–	18:34:54.6			4	294	176	139	0.506	0.389
Leipzig	51 19.0	12 20.0	–	18:32:29.7			2	297	176	140	0.469	0.349
Mannheim	49 29.0	8 29.0	–	18:36:12.9			2	295	176	138	0.523	0.407
Munich	48 08.0	11 35.0	571	18:36:35.9			0	298	176	138	0.528	0.412
Nurnberg	49 27.0	11 04.0	344	18:35:15.8			1	297	176	138	0.508	0.391
Stuttgart	48 46.0	9 11.0	–	18:36:49.5			2	296	176	138	0.531	0.416
Wiesbaden	50 05.0	8 14.0	–	18:35:33.7			3	295	176	138	0.514	0.397
Wuppertal	51 16.0	7 11.0	–	18:34:25.1			4	294	176	139	0.499	0.382
HUNGARY												
Budapest	47 30.0	19 05.0	129	18:08 Set			0	298	–	–	0.330	0.211
IRELAND												
Dublin	53 20.0	-6-15.0	51	18:34:07.5			12	283	175	138	0.539	0.425
ITALY												
Bologna	44 29.0	11 20.0	–	18:27 Set			0	296	–	–	0.535	0.420
Catania	37 30.0	15 06.0	–	17:55 Set			0	293	–	–	0.105	0.039
Florence	43 46.0	11 15.0	–	18:25 Set			0	295	–	–	0.525	0.409
Genova	44 25.0	8 57.0	104	18:38 Set			0	296	–	–	0.608	0.502
Milano	45 28.0	9 12.0	–	18:40:39.8			0	297	177	135	0.591	0.483
Napoli	40 51.0	14 17.0	27	18:07 Set			0	294	–	–	0.300	0.185
Palermo	38 07.0	13 21.0	116	18:05 Set			0	293	–	–	0.258	0.148
Rome	41 54.0	12 29.0	124	18:18 Set			0	295	–	–	0.446	0.325
Torino	45 03.0	7 40.0	–	18:41:46.5			0	296	177	135	0.609	0.503
LATVIA												
Riga	56 57.0	24 06.0	–	18:20:33.8			0	305	175	147	0.321	0.203

LOCAL CIRCUMSTANCES DURING THE ANNULAR SOLAR ECLIPSE OF 10 MAY 1994 FOR EUROPE

Location Name	First Contact				Second Contact				Third Contact				Fourth Contact			
	U.T. h m s	Alt °	P °	V °	U.T. h m s	Alt °	P °	V °	U.T. h m s	Alt °	P °	V °	U.T. h m s	Alt °	P °	V °
ANDORRA																
Andorra la Ve..	17:44:20.1	13	251	202									-			
AUSTRIA																
Vienna	17:40:56.4	5	237	196									-			
BELGIUM																
Antwerp	17:37:32.9	14	238	198									-			
Brussels	17:37:50.4	14	239	198									-			
Liege	17:38:16.5	13	238	198									-			
BYELARUS																
Minsk	17:36:45.2	2	226	192									-			
CZECHOSLOVAKIA																
Ostrava	17:39:49.2	5	234	195									-			
Prague	17:39:42.2	7	236	196									-			
DENMARK																
Copenhagen	17:35:36.8	11	229	195									-			
ESTONIA																
Tallinn	17:33:55.1	6	220	191									-			
FINLAND																
Helsinki	17:33:29.3	6	219	191									-			
FRANCE																
Bordeaux	17:41:43.6	15	249	201									-			
Lille	17:37:41.5	15	239	198									-			
Lyon	17:42:11.2	12	245	200									-			
Marseille	17:44:13.2	10	248	200									-			
Paris	17:38:59.0	15	242	199									-			
Toulouse	17:43:19.5	13	249	201									-			
GERMANY																
Aachen	17:38:15.6	13	238	197									-			
Berlin	17:37:57.1	9	233	195									-			
Bielefeld	17:37:44.0	12	236	197									-			
Bonn	17:38:29.2	12	238	197									-			
Bremen	17:36:58.8	12	234	196									-			
Dortmund	17:37:56.5	12	237	197									-			
Dresden	17:39:00.1	8	235	196									-			
Duisburg	17:37:53.4	13	237	197									-			
Dusseldorf	17:38:03.9	12	237	197									-			
Essen	17:37:05.7	12	235	196									-			
Frankfurt	17:39:12.9	11	238	197									-			
Hamburg	17:36:48.1	11	233	196									-			
Hannover	17:37:37.7	11	235	196									-			
Koln	17:38:18.7	12	238	197									-			
Leipzig	17:38:43.2	9	235	196									-			
Mannheim	17:39:40.5	11	239	197									-			
Munich	17:40:58.6	8	239	197									-			
Nurnberg	17:39:58.7	9	238	197									-			
Stuttgart	17:40:18.4	10	239	197									-			
Wiesbaden	17:39:10.7	11	238	197									-			
Wuppertal	17:38:05.2	12	237	197									.			
HUNGARY																
Budapest	17:41:12.1	3	237	196									-			
IRELAND																
Dublin	17:31:50.4	21	239	200									19:31:36.9	4	112	77
ITALY																
Bologna	17:43:36.8	7	244	199									-			
Catania	17:48:07.9	1	252	201									-			
Florence	17:44:08.3	7	245	199									-			
Genova	17:43:36.6	8	245	199									-			
Milano	17:42:49.1	9	244	199									-			
Napoli	17:45:59.8	3	248	200									-			
Palermo	17:48:01.1	3	252	201									-			
Rome	17:45:26.3	5	247	200									-			
Torino	17:43:01.8	10	245	199									-			
LATVIA																
Riga	17:35:21.8	5	223	192									.			

CIRCUMSTANCES AT MAXIMUM ECLIPSE ON 10 MAY 1994
FOR EUROPE

Location Name	Latitude ° '	Longitude ° '	Elev. m	U.T. h m s	Umbral Durat. m s	Path Width km	Sun Alt °	Sun Az. °	P °	V °	Eclipse Mag.	Eclipse Obs.
LIECHTENSTEIN												
Vaduz	47 09.0	9 31.0	–	18:38:36.2			1	297	176	136	0.558	0.446
LITHUANIA												
Vilnius	54 40.0	25 26.0	–	18:06 Set			0	303	–	–	0.291	0.177
LUXEMBOURG												
Luxembourg	49 36.0	6 09.0	360	18:36:52.7			4	294	176	137	0.535	0.420
MALTA												
Valletta	35 54.0	14 31.0	76	17:56 Set			0	293	–	–	0.096	0.035
MONACO												
Monaco	43 44.0	7 25.0	59	18:42 Set			0	296	–	–	0.635	0.533
NETHERLANDS												
Amsterdam	52 22.0	4 54.0	2	18:33:39.0			6	292	176	139	0.493	0.375
Rotterdam	51 55.0	4 28.0	–	18:34:22.0			6	292	176	139	0.504	0.386
S'Gravenhage	52 06.0	4 18.0	–	18:34:10.0			6	292	176	139	0.501	0.384
Utrecht	52 05.0	5 08.0	–	18:33:57.8			5	292	176	139	0.497	0.379
NORWAY												
Oslo	59 55.0	10 45.0	101	18:21:38.0			7	293	175	146	0.345	0.225
POLAND												
Gdansk	54 23.0	18 40.0	12	18:26:06.0			1	301	175	144	0.385	0.264
Krakow	50 03.0	19 58.0	237	18:14 Set			0	300	–	–	0.378	0.257
Lodz	51 46.0	19 30.0	–	18:18 Set			0	300	–	–	0.393	0.272
Poznan	52 25.0	16 55.0	–	18:29:13.4			0	300	176	142	0.425	0.304
Warsaw	52 15.0	21 00.0	96	18:16 Set			0	301	–	–	0.377	0.256
Wroclaw	51 06.0	17 00.0	158	18:29 Set			0	300	–	–	0.444	0.324
PORTUGAL												
Lisbon	38 43.0	-9-08.0	103	18:53:46.6			7	287	177	125	0.855	0.792
Porto	41 10.0	-8-36.0	–	18:50:47.1			8	287	177	127	0.798	0.724
SAN MARINO												
San Marino	43 55.0	12 28.0	–	18:21 Set			0	295	–	–	0.479	0.360
SPAIN												
Barcelona	41 23.0	2 11.0	102	18:47:59.2			1	293	177	130	0.718	0.629
Bilbao	43 15.0	-2-58.0	–	18:47:17.3			5	289	177	130	0.715	0.626
Madrid	40 24.0	-3-41.0	718	18:50:50.5			4	290	177	128	0.780	0.702
Malaga	36 34.0	-4-25.0	–	18:55:16.0			3	290	177	125	0.869	0.807
Seville	37 23.0	-5-59.0	32	18:54:45.3			4	289	177	125	0.862	0.799
Valencia	39 28.0	0-22.0	26	18:50:57.5			2	292	177	128	0.776	0.698
Zaragoza	41 38.0	0-53.0	–	18:48:40.6			3	291	177	130	0.734	0.648
SWEDEN												
Goteborg	57 43.0	11 58.0	18	18:24:21.1			5	295	175	144	0.371	0.250
Stockholm	59 20.0	18 03.0	48	18:20:07.7			4	299	175	147	0.318	0.201
SWITZERLAND												
Basel	47 33.0	7 35.0	–	18:38:53.5			2	295	176	136	0.563	0.451
Bern	46 57.0	7 26.0	616	18:39:40.1			1	295	176	136	0.575	0.464
Zurich	47 23.0	8 32.0	531	18:38:43.9			1	296	176	136	0.560	0.448
UKRAINE												
L'vov	49 50.0	24 00.0	321	17:57 Set			0	300	–	–	0.223	0.119
UNITED KINGDOM												
Belfast	54 35.0	-5-55.0	19	18:32:15.0			13	283	175	139	0.514	0.398
Birmingham	52 29.0	-1-55.0	176	18:34:55.1			9	287	176	138	0.530	0.415
Bristol	51 27.0	-2-35.0	–	18:36:28.2			9	287	176	137	0.553	0.440
Cardiff	51 29.0	-3-13.0	67	18:36:30.7			10	286	176	137	0.556	0.443
Coventry	52 25.0	-1-30.0	–	18:34:56.9			9	287	176	138	0.529	0.414
Edinburgh	55 57.0	-3-13.0	145	18:30:00.9			12	284	175	140	0.476	0.357
Glasgow	55 53.0	-5-15.0	–	18:30:15.5			13	283	175	140	0.488	0.370
Leeds	53 50.0	-1-35.0	–	18:32:56.4			10	287	175	139	0.504	0.387
Liverpool	53 25.0	-2-55.0	65	18:33:42.9			10	286	175	138	0.519	0.403
London	51 30.0	0-10.0	49	18:35:59.8			8	289	176	137	0.538	0.424
Manchester	53 28.0	-2-15.0	–	18:33:33.5			10	286	175	138	0.514	0.398
Middlesbrough	54 35.0	-1-14.0	–	18:31:48.4			10	286	175	139	0.489	0.371
Newcastle	52 26.0	-3-06.0	–	18:35:08.9			10	286	176	137	0.538	0.424
Nottingham	52 58.0	-1-10.0	–	18:34:07.1			9	287	176	138	0.517	0.401
Sheffield	53 23.0	-1-28.0	–	18:33:34.2			10	287	175	138	0.512	0.395
YUGOSLAVIA												
Belgrade	44 50.0	20 30.0	149	17:54 Set			0	296	–	–	0.160	0.073
Zagreb	45 48.0	15 58.0	–	18:12 Set			0	296	–	–	0.382	0.260

Table 12b
LOCAL CIRCUMSTANCES DURING THE ANNULAR SOLAR ECLIPSE OF 10 MAY 1994
FOR EUROPE

Location Name	First Contact				Second Contact				Third Contact				Fourth Contact			
	U.T. h m s	Alt °	P °	V °	U.T. h m s	Alt °	P °	V °	U.T. h m s	Alt °	P °	V °	U.T. h m s	Alt °	P °	V °
LIECHTENSTEIN																
Vaduz	17:41:33.8	9	241	198												
LITHUANIA																
Vilnius	17:36:33.9	3	225	192									–			
LUXEMBOURG																
Luxembourg	17:39:13.5	12	240	198												
MALTA																
Valletta	17:49:20.5	1	254	202									–			
MONACO																
Monaco	17:44:03.1	9	247	200												
NETHERLANDS																
Amsterdam	17:36:43.8	14	236	197									–			
Rotterdam	17:36:59.4	14	237	198												
S'Gravenhage	17:36:48.0	14	237	198									–			
Utrecht	17:37:00.8	14	237	197												
NORWAY																
Oslo	17:32:09.6	13	225	194									19:08:42.2	2	125	99
POLAND																
Gdansk	17:36:56.5	7	229	194									–			
Krakow	17:39:35.6	4	233	195												
Lodz	17:38:34.4	5	231	194												
Poznan	17:38:11.1	7	232	195												
Warsaw	17:38:13.1	4	230	194									–			
Wroclaw	17:39:02.9	6	233	195												
PORTUGAL																
Lisbon	17:44:40.9	20	260	206												
Porto	17:42:19.0	21	257	205									–			
SAN MARINO																
San Marino	17:44:00.1	6	244	199									–			
SPAIN																
Barcelona	17:45:25.5	12	252	202									–			
Bilbao	17:42:27.9	17	252	202												
Madrid	17:45:00.0	16	256	204												
Malaga	17:48:38.8	15	262	206									–			
Seville	17:47:19.9	17	261	206									–			
Valencia	17:46:43.0	13	256	203									–			
Zaragoza	17:44:34.6	15	253	203									–			
SWEDEN																
Goteborg	17:34:01.4	12	227	194									19:12:09.5	0	123	96
Stockholm	17:33:39.5	9	223	192									–			
SWITZERLAND																
Basel	17:41:03.6	11	242	198									–			
Bern	17:41:31.4	10	243	199									–			
Zurich	17:41:18.4	10	242	198									–			
UKRAINE																
L'vov	17:39:17.8	2	232	194									–			
UNITED KINGDOM																
Belfast	17:30:55.0	21	237	200									19:29:01.6	5	113	79
Birmingham	17:34:30.0	18	239	199									19:30:54.1	2	113	78
Bristol	17:35:08.1	19	240	200									19:33:08.9	1	111	76
Cardiff	17:34:51.0	19	241	200									19:33:26.8	2	111	75
Coventry	17:34:42.9	18	239	199									19:30:47.0	2	113	78
Edinburgh	17:30:59.9	20	235	198									19:24:57.9	5	116	84
Glasgow	17:30:07.5	21	235	199									19:26:06.6	6	115	82
Leeds	17:33:28.0	19	237	199									19:28:12.2	3	114	81
Liverpool	17:33:17.3	19	238	199									19:29:43.7	3	113	79
London	17:35:59.1	17	239	199									19:31:38.0	1	112	77
Manchester	17:33:31.0	19	238	199									19:29:16.0	3	114	80
Middlesbrough	17:32:58.0	18	236	198									19:26:34.8	3	115	82
Newcastle	17:34:03.8	19	239	199									19:31:39.5	2	112	77
Nottingham	17:34:22.1	18	238	199									19:29:35.0	2	113	79
Sheffield	17:33:53.8	18	237	199									19:28:58.9	2	114	80
YUGOSLAVIA																
Belgrade	17:42:35.3	1	240	197									–			
Zagreb	17:42:32.0	4	241	197									–			

Table 13a
CIRCUMSTANCES AT MAXIMUM ECLIPSE ON 10 MAY 1994
FOR THE NORTH ATLANTIC

Location Name	Latitude ° ′	Longitude ° ′	Elev. m	U.T. h m s	Umbral Durat. m s	Path Width km	Sun Alt °	Sun Az. °	P °	V °	Eclipse Mag.	Eclipse Obs.
AZORES												
Angra do Heroismo	38 39.0	-27-13.0	—	18:51:11.2	4 19.5	278	21	276	356	301	0.935	0.875
Horta	38 32.0	-28-38.0	—	18:50:38.7	2 20.1	277	23	275	356	301	0.936	0.875
Ponta Delgada	37 44.0	-25-40.0	39	18:53:00.9	2 24.7	282	20	275	356	301	0.935	0.874
Santa Cruz da Gra..	39 05.0	-28-01.0	—	18:50:14.8	4 44.1	277	22	275	356	302	0.935	0.875
Sao Mateus	38 26.0	-28-27.0	—	18:50:52.0	1 54.6	278	22	275	356	301	0.935	0.875
BERMUDA												
Hamilton	32 17.0	-64-46.0	50	17:59:03.4			62	244	347	294	0.625	0.524
CANARY ISLANDS												
Arrecife	28 57.0	-13-32.0	—	19:04:33.4			6	287	358	297	0.812	0.740
Las Palmas G.Cana..	28 07.0	-15-28.0	7	19:05:27.0			7	287	358	295	0.775	0.697
Santa Cruz la Pal..	28 41.0	-17-45.0	—	19:04:53.6			9	285	358	295	0.772	0.693
Santa Cruz Teneri..	28 25.0	-16-16.0	—	19:05:11.0			8	286	358	295	0.777	0.699
CAPE VERDE												
Praia	14 55.0	-23-31.0	37	19:14:47.5			8	286	359	282	0.358	0.238
GREENLAND												
Godthab	64 11.0	-51-44.0	66	17:58:38.5			37	228	170	150	0.480	0.361
ICELAND												
Akureyri	65 44.0	-18-08.0	—	18:13:04.5			22	265	173	147	0.374	0.254
Reykjavik	64 09.0	-21-51.0	30	18:14:50.3			23	262	173	146	0.413	0.293

Table 14a
CIRCUMSTANCES AT MAXIMUM ECLIPSE ON 10 MAY 1994
FOR AFRICA

Location Name	Latitude ° ′	Longitude ° ′	Elev. m	U.T. h m s	Umbral Durat. m s	Path Width km	Sun Alt °	Sun Az. °	P °	V °	Eclipse Mag.	Eclipse Obs.
ALGERIA												
Algiers	36 47.0	3 03.0	64	18:43 Set			0	293	—	—	0.749	0.666
Annaba	36 54.0	7 46.0	22	18:24 Set			0	293	—	—	0.502	0.385
Constantine	36 22.0	6 37.0	—	18:26 Set			0	293	—	—	0.534	0.419
Wahran	35 43.0	0 43.0	—	18:49 Set			0	292	—	—	0.818	0.747
BURKINA FASO												
Bobo-Dioulasso	11 12.0	-4-18.0	—	18:29 Set			0	288	—	—	0.103	0.038
GUINEA												
Conakry	9 31.0	-13-43.0	8	19:05 Set			0	288	—	—	0.273	0.161
GUINEA-BISSAU												
Bissau	11 51.0	-15-35.0	—	19:16:27.0			0	288	359	282	0.344	0.224
LIBERIA												
Monrovia	6 18.0	-10-47.0	25	18:49 Set			0	288	—	—	0.126	0.051
MALI												
Bamako	12 39.0	-8 00.0	366	18:49 Set			0	288	—	—	0.308	0.192
MAURITANIA												
Nouakchott	18 06.0	-15-57.0	23	19:13:16.2			3	288	359	287	0.510	0.393
MOROCCO												
Agadir	30 26.0	-9-36.0	—	19:02:36.9			3	289	358	299	0.879	0.818
Beni-Mellal	32 22.0	-6-29.0	—	19:00:04.3	4 4.1	303	2	290	358	301	0.930	0.866
Casablanca	33 35.0	-7-30.0	54	18:59:06.1	4 33.0	298	3	289	178	122	0.931	0.866
Fes	34 05.0	-4-57.0	—	18:57:58.5			2	290	178	123	0.929	0.865
Kenitra	34 16.0	-6-40.0	—	18:58:13.2	2 51.1	295	3	290	178	123	0.931	0.866
Khouribga	32 54.0	-6-57.0	—	18:59:39.8	4 29.5	301	3	290	358	301	0.931	0.866
Marrakech	31 38.0	-8 00.0	495	19:01:08.9			3	289	358	300	0.920	0.860
Meknes	33 53.0	-5-37.0	—	18:58:21.1	3 0.4	295	2	290	178	123	0.930	0.866
Oujda	34 41.0	-1-45.0	—	18:56:23.6			0	292	178	124	0.891	0.832
Rabat	34 02.0	-6-51.0	70	18:58:30.0	3 42.7	296	3	289	178	122	0.931	0.866
Safi	32 20.0	-9-17.0	—	19:00:42.6			4	289	358	300	0.927	0.865
Tangier	35 48.0	-5-45.0	78	18:56:24.7			3	290	178	124	0.896	0.837
Tetouan	35 34.0	-5-23.0	—	18:56:34.0			3	290	178	124	0.899	0.840
SENEGAL												
Dakar	14 40.0	-17-26.0	43	19:15:19.5			3	288	359	283	0.405	0.283
SIERRA LEONE												
Freetown	8 30.0	-13-15.0	30	19:02 Set			0	288	—	—	0.238	0.131
TUNISIA												
Sfax	34 44.0	10 46.0	—	18:07 Set			0	292	—	—	0.238	0.131
Tunis	36 48.0	10 11.0	71	18:15 Set			0	293	—	—	0.375	0.255

LOCAL CIRCUMSTANCES DURING THE ANNULAR SOLAR ECLIPSE OF 10 MAY 1994
FOR THE NORTH ATLANTIC

Location Name	First Contact U.T. h m s	Alt	P	V	Second Contact U.T. h m s	Alt	P	V	Third Contact U.T. h m s	Alt	P	V	Fourth Contact U.T. h m s	Alt	P	V
AZORES																
Angra do Hero..	17:30:19.6	37	267	212	18:49:02.1	22	297	242	18:53:21.6	21	56	1	20:01:31.6	8	86	34
Horta	17:28:44.1	39	267	213	18:49:30.0	23	329	274	18:51:50.1	22	24	329	20:01:44.0	9	85	33
Ponta Delgada	17:33:17.8	35	268	212	18:51:49.7	20	328	272	18:54:14.4	19	26	330	20:02:25.2	6	86	33
Santa Cruz da..	17:28:48.4	38	266	212	18:47:53.2	23	287	232	18:52:37.3	22	67	12	20:01:03.8	9	86	35
Sao Mateus	17:29:05.6	38	268	213	18:49:56.2	23	335	279	18:51:50.8	22	19	324	20:01:50.6	9	85	33
BERMUDA																
Hamilton	16:10:27.3	75	277	281									19:37:56.5	42	61	359
CANARY ISLANDS																
Arrecife	17:54:34.9	20	277	212									-			
Las Palmas G....	17:54:53.4	22	279	213									-			
Santa Cruz la..	17:52:51.9	24	279	213									-			
Santa Cruz Te..	17:54:03.9	23	279	213									-			
CAPE VERDE																
Praia	18:16:12.7	22	308	227									-			
GREENLAND																
Godthab	16:46:49.3	41	230	218									19:06:07.0	31	112	87
ICELAND																
Akureyri	17:15:18.2	28	225	201									19:07:23.8	16	121	96
Reykjavik	17:13:42.7	30	228	203									19:11:59.2	17	119	91

LOCAL CIRCUMSTANCES DURING THE ANNULAR SOLAR ECLIPSE OF 10 MAY 1994
FOR AFRICA

Location Name	First Contact U.T. h m s	Alt	P	V	Second Contact U.T. h m s	Alt	P	V	Third Contact U.T. h m s	Alt	P	V	Fourth Contact U.T. h m s	Alt	P	V
ALGERIA																
Algiers	17:49:37.1	10	258	204									-			
Annaba	17:49:31.3	6	256	203									-			
Constantine	17:50:00.8	7	257	203									-			
Wahran	17:50:25.0	11	261	205									-			
BURKINA FASO																
Bobo-Dioulasso	18:20:36.3	2	303	224									-			
GUINEA																
Conakry	18:28:03.2	8	313	230									-			
GUINEA-BISSAU																
Bissau	18:23:20.5	12	309	228									-			
LIBERIA																
Monrovia	18:34:11.7	3	319	234									-			
MALI																
Bamako	18:19:40.0	6	302	224									-			
MAURITANIA																
Nouakchott	18:10:51.2	17	297	221									-			
MOROCCO																
Agadir	17:54:04.7	17	273	210									-			
Beni-Mellal	17:52:39.6	16	269	208	18:58:02.3	2	295	238	19:02:06.4	2	61	4	-			
Casablanca	17:51:00.1	17	267	208	18:56:49.6	4	260	204	19:01:22.5	3	96	40	-			
Fes	17:51:08.0	15	266	207									-			
Kenitra	17:50:29.0	16	266	207	18:56:47.7	3	216	161	18:59:38.7	3	139	84	-			
Khouribga	17:51:56.1	16	268	208	18:57:25.0	3	279	223	19:01:54.5	2	76	20	..			
Marrakech	17:53:06.8	17	270	209									..			
Meknes	17:51:11.4	15	266	207	18:56:51.0	3	219	164	18:59:51.4	2	136	81	-			
Oujda	17:51:07.2	12	263	206									-			
Rabat	17:50:41.4	17	266	207	18:56:38.6	4	232	176	19:00:21.4	3	124	68	-			
Safi	17:51:52.8	18	270	209									..			
Tangier	17:49:05.1	16	263	206									-			
Tetouan	17:49:26.2	16	264	206									-			
SENEGAL																
Dakar	18:17:27.4	16	304	225									-			
SIERRA LEONE																
Freetown	18:30:15.0	7	315	232									-			
TUNISIA																
Sfax	17:50:54.0	3	258	203									.			
Tunis	17:49:24.1	4	255	202									.			

Table 15

CLIMATE STATISTICS DURING MAY
FOR SELECTED STATIONS WITHIN THE UMBRAL PATH

Station	Mean High Temp. °F	Mean Low Temp. °F	Prevailing Wind	Days with ≤3/10ths cloud and good visibility	Sunshine hours E = estimated	Days with Rain
Mexico						
Puerto Cortes	69	62	N	21.7	-	1.0
Hermosillo	96	59	-	24.0	-	0.5
Guaymas	70	58	-	24.7	310	0.5
Chihuahua	87	58	-	21.7	284	0.9
Nuevo Casas Grandes	91	62	-	20.6	-	0.9
United States						
Bisbee-Douglas, Arizona	85	49	-	19.7	388E	0.3
Alamagordo, New Mexico	85	55	-	16.7	360E	0.5
Las Cruces, New Mexico	83	60	-	17.4	370E	1.1
Deming, New Mexico	85	49	-	-	380E	0.6
Roswell, New Mexico	85	55	S	15.6	330	3.2
El Paso, Texas	87	57	WSW	17.6	373	1.1
Lubbock, Texas	83	55	S	12.5	315E	5.8
Childress, Texas	81	57	-	11.3	305E	7.1
Amarillo, Texas	79	52	S	12.3	305	5.9
Altus, Oklahoma	82	60	SSE	9.9	300E	6.8
Oklahoma City, Oklahoma	79	58	S	10.5	290	7.3
Wichita, Kansas	77	55	S	8.2	291	6.3
Kansas City, Kansas	74	54	S	7.3	278	7.2
Jefferson City, Missouri	75	54	-	6.9	280E	7.1
Springfield, Illinois	74	53	S	7.2	282	7.3
Toledo, Ohio	71	47	ENE	6.2	263	6.6
Detroit, Michigan	70	47	W	5.1	263	5.7
Cleveland, Ohio	69	48	N	7.4	274	6.7
Rochester, New York	68	46	WSW	6.5	274	6.3
Burlington, Vermont	67	44	S	5.2	244	6.3
Portland, Maine	63	43	S	6.8	268	6.6
Augusta, Maine	65	43	-	4.5	290E	6.7
Canada						
Toronto, Ontario	65	43	N	5.8	233	7.8
St John, New Brunswick	58	38	SSW	7.0	203	8.1
Halifax, Nova Scotia	58	39	S	-	207	-
Azores, Portugal						
Corvo, Flores	65	59	-	3.3	162	6.6
Horta, Ilha do Pico	67	57	-	5.8	177	5.9
Lajes, Ilha Terceira	66	57	NW	0.7	160	4.1
Ponta Delgada, Ilha de Sao Miguel	67	56	-	3.2	158	5.0
Morocco						
Marrakesh	84	57	W	29.3	288	1.5
Casablanca	72	56	N	12.8	292	2.2
Ifrane	66	41	-	-	244	
Midelt	73	48	-	-	303	
Kenitra	75	58	NW	12.5	298	2.0

ANNULAR SOLAR ECLIPSE OF 10 MAY 1994

MAPS
OF THE
UMBRAL PATH

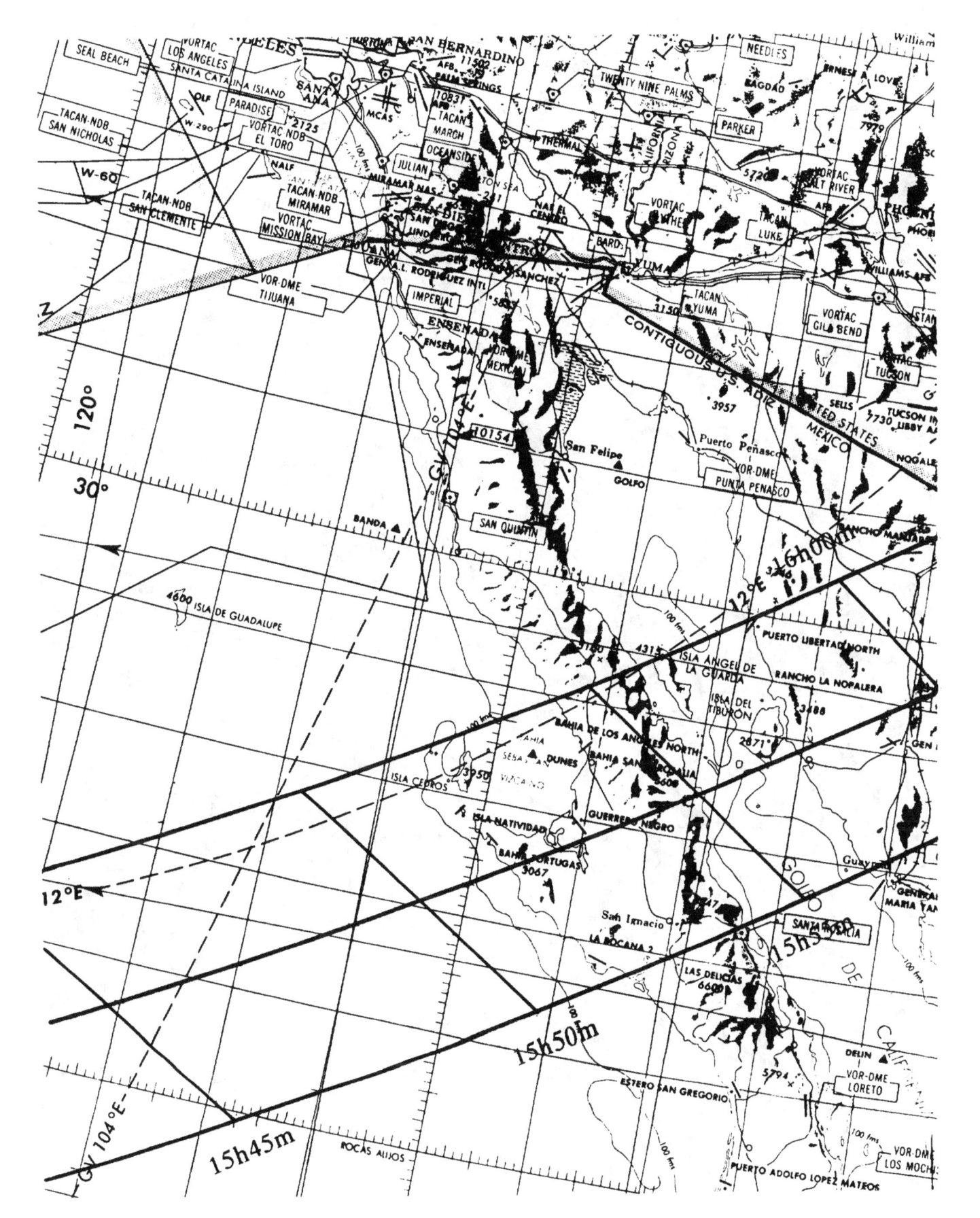

79

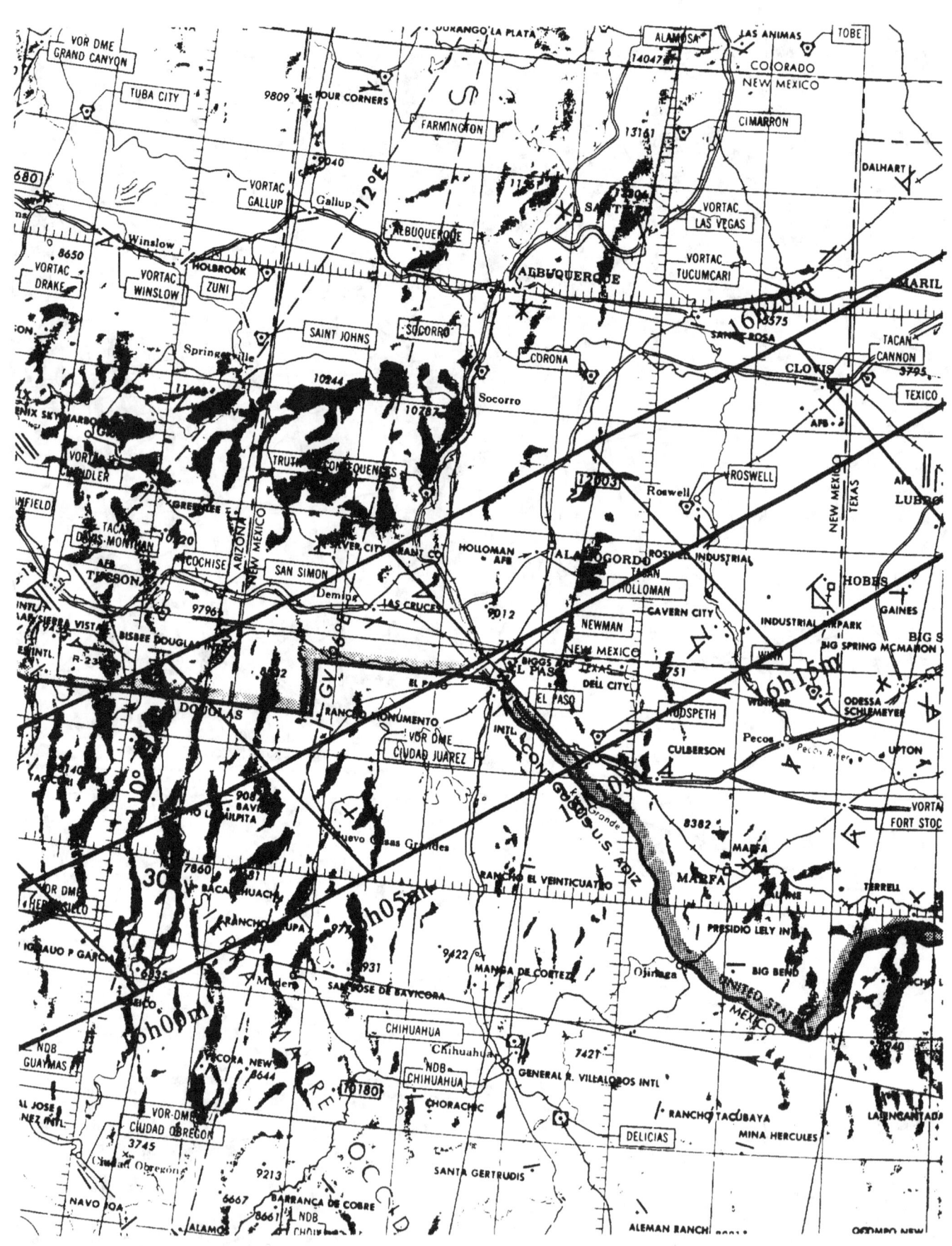

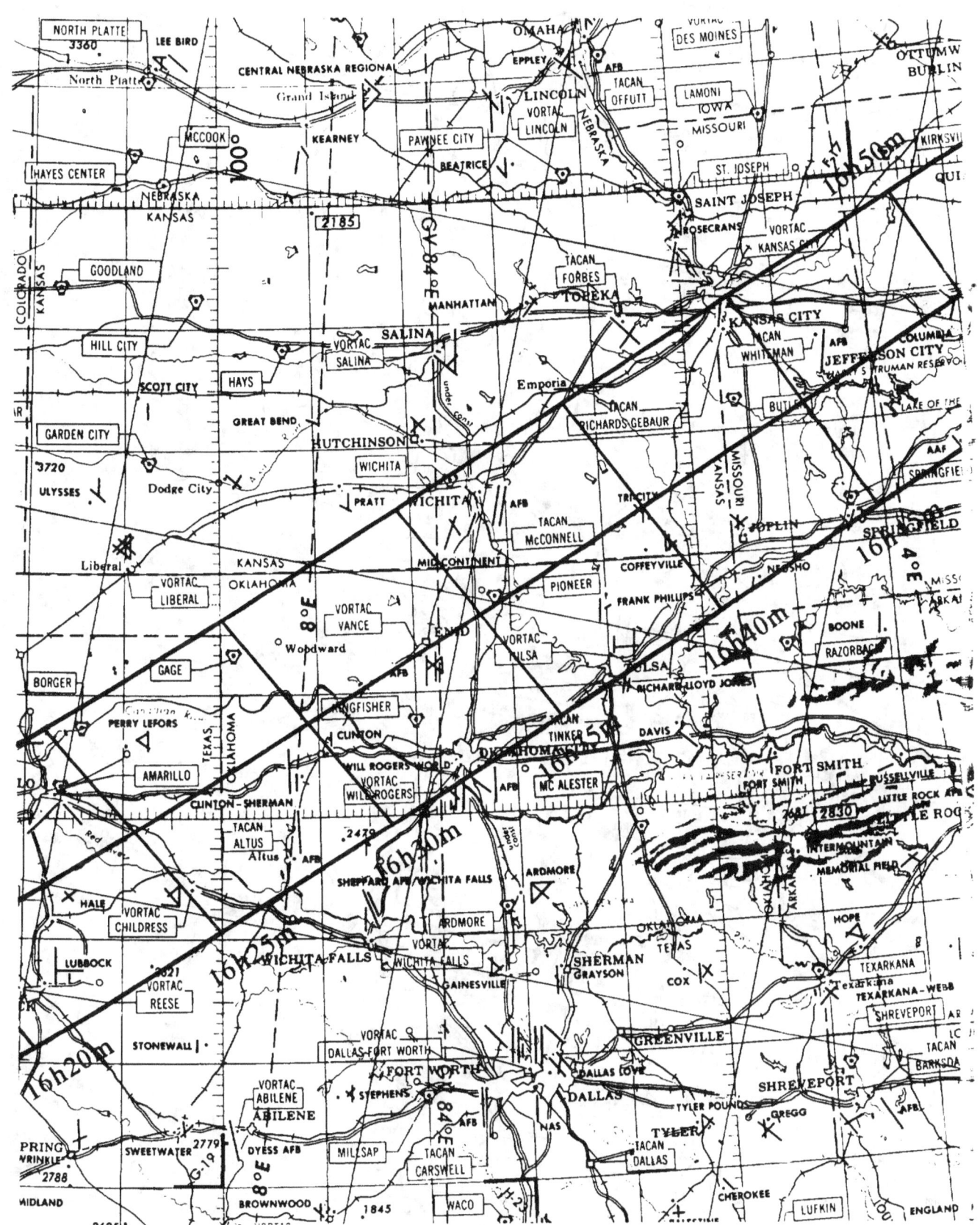

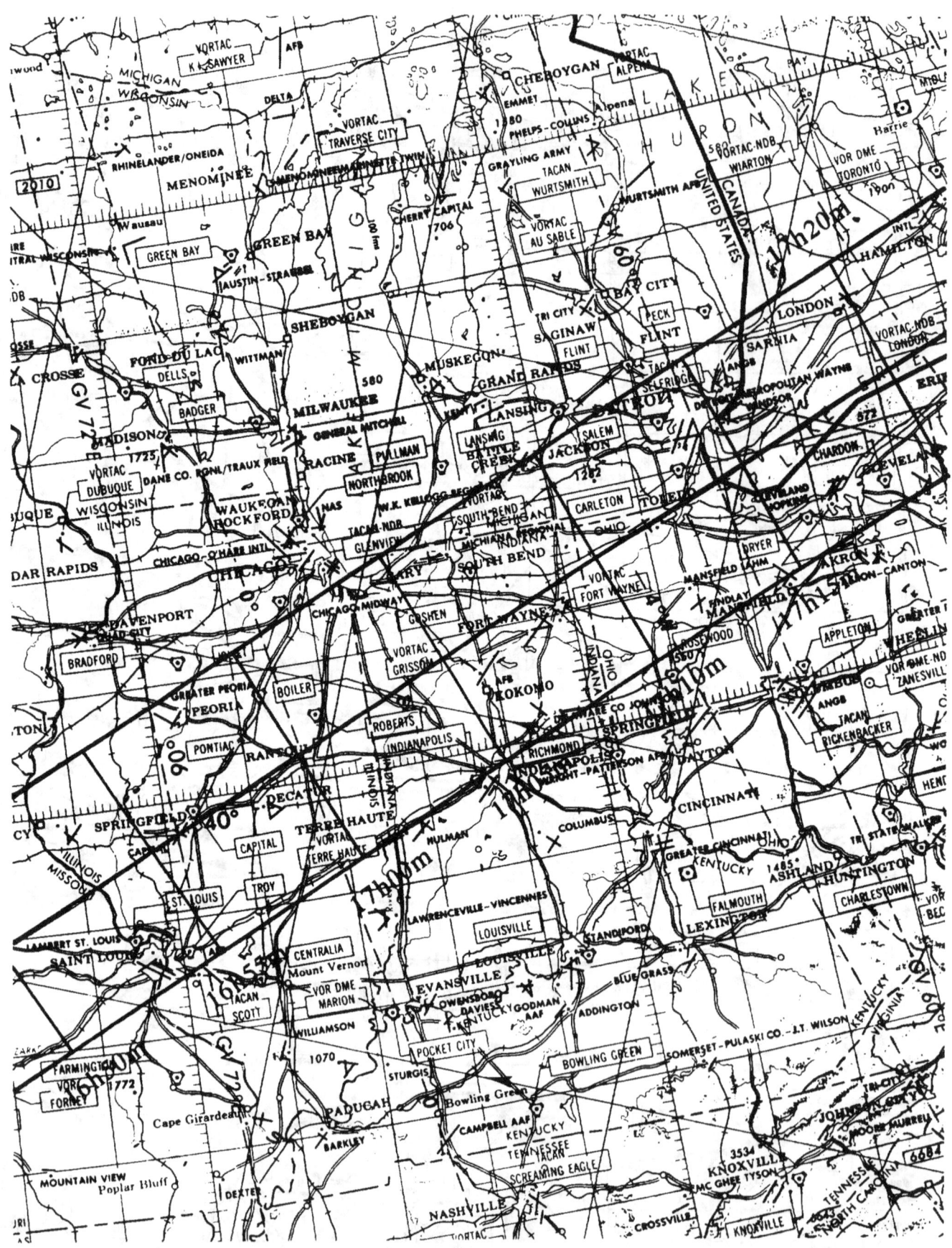

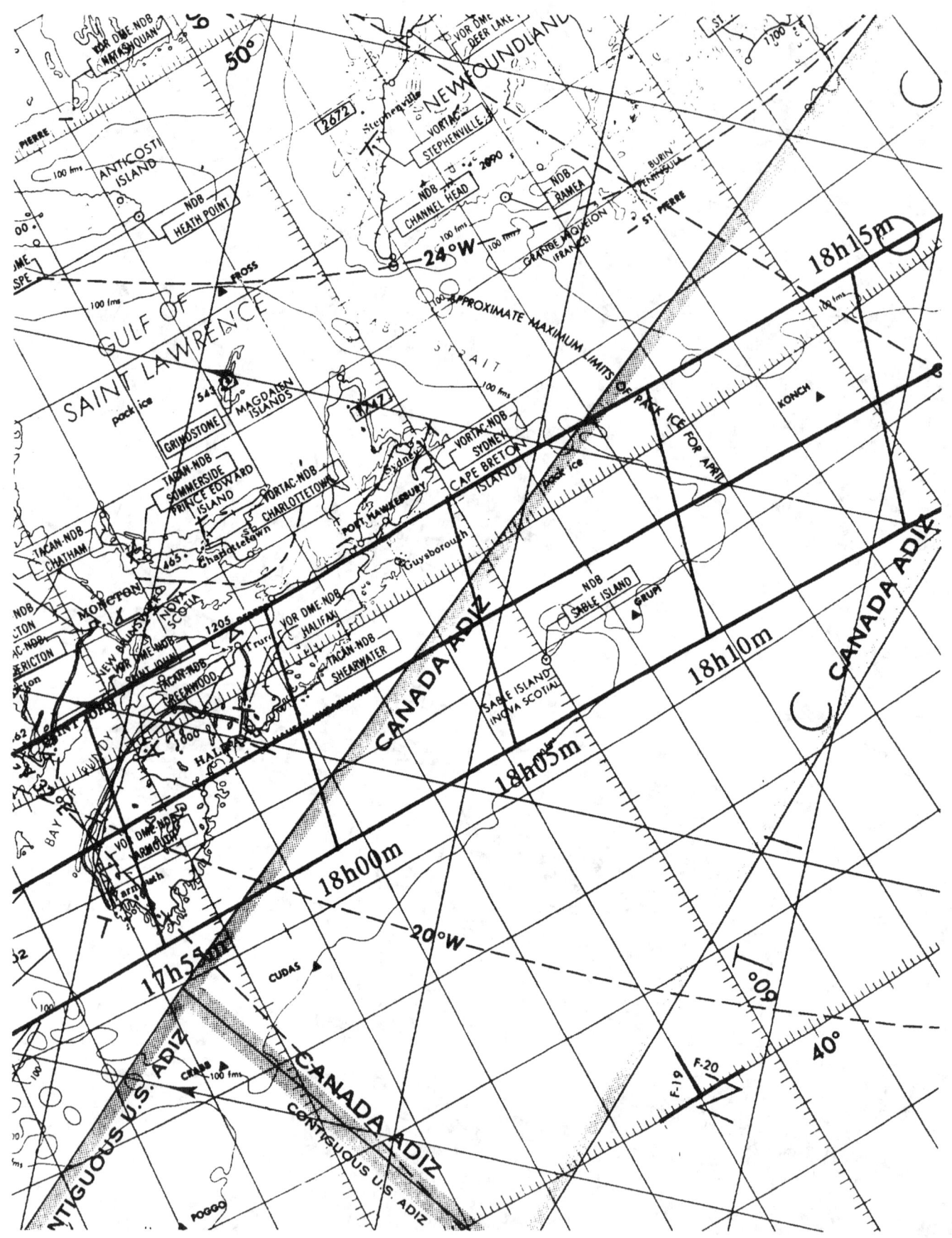

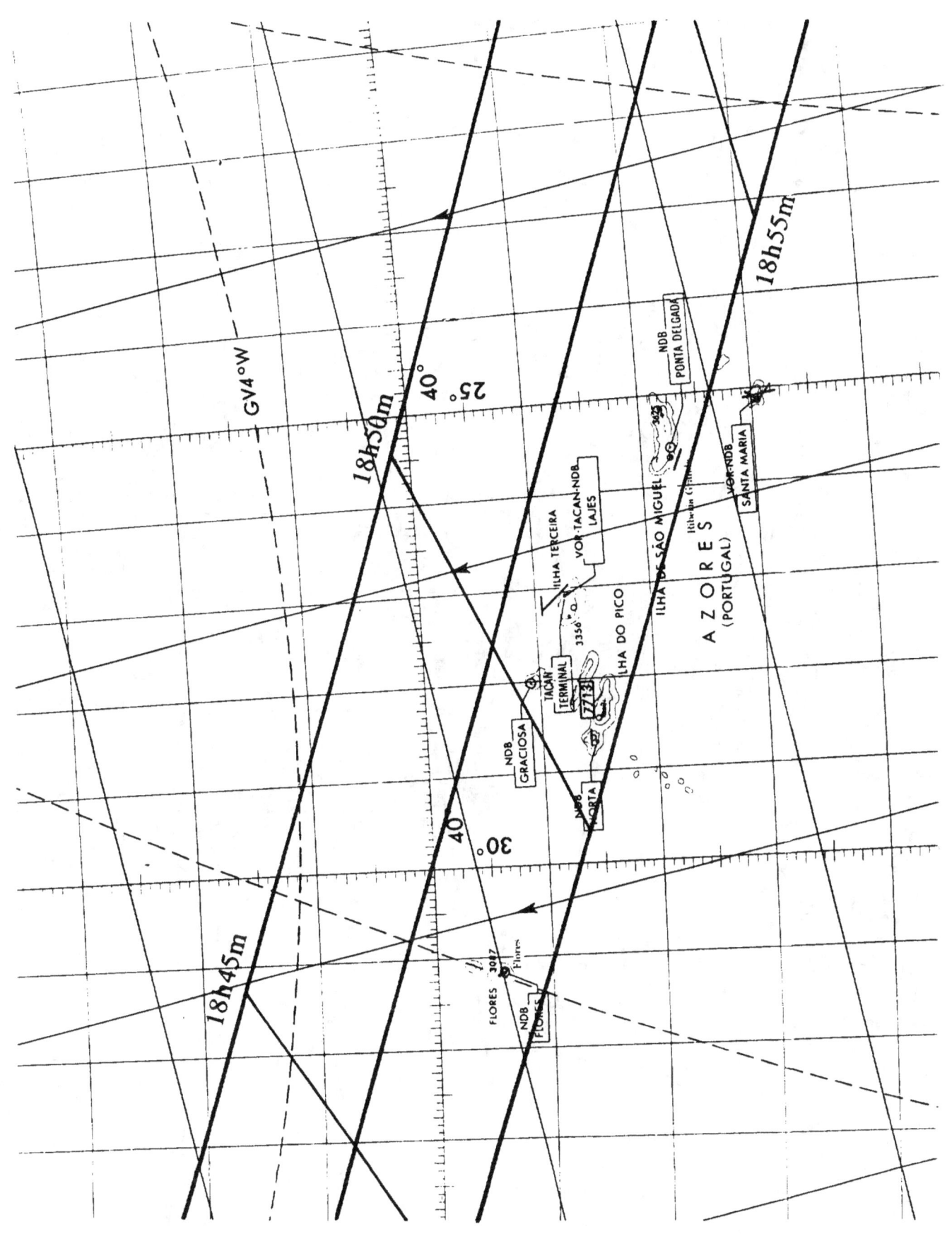

85

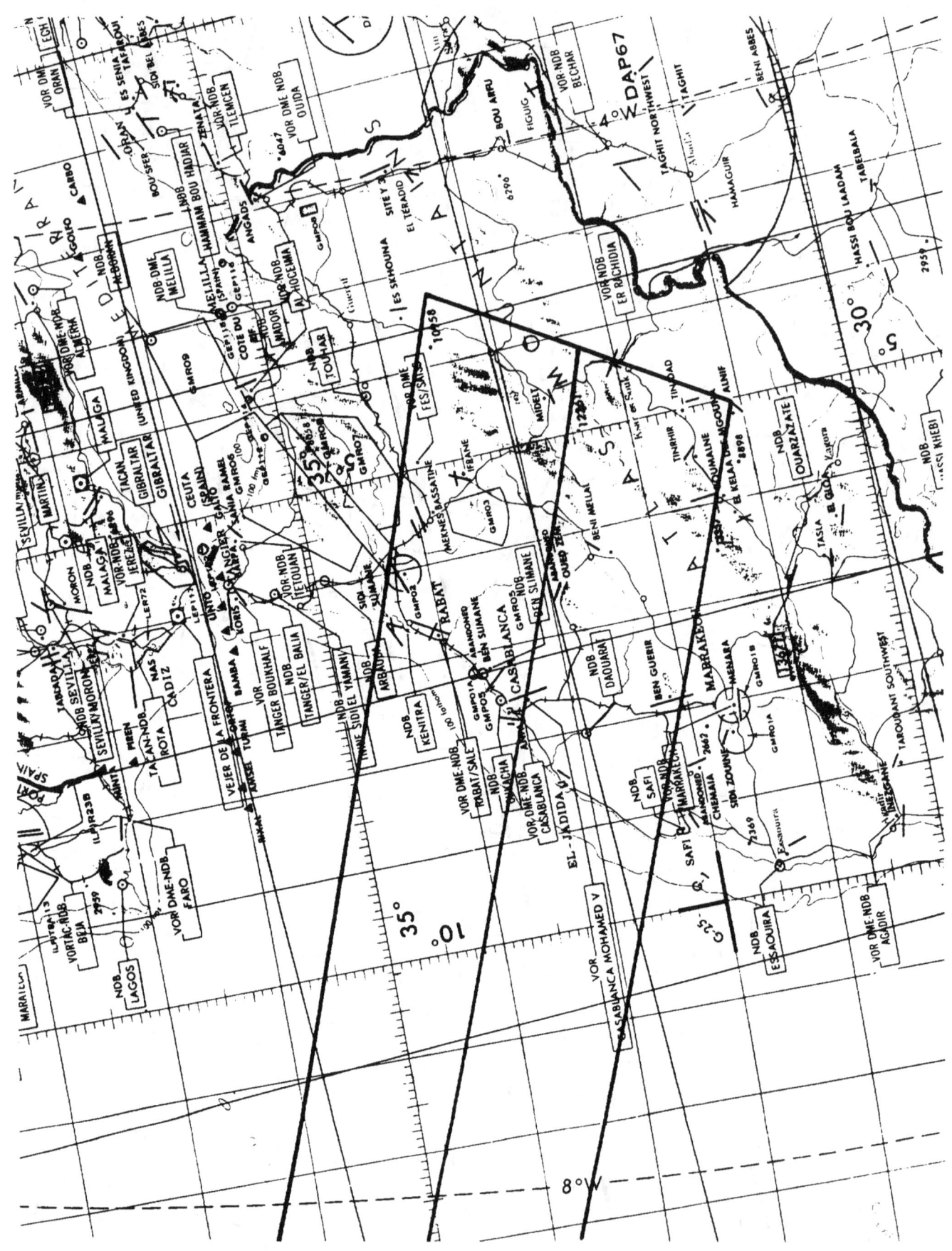